Farmacology

ALSO BY DAPHNE MILLER, M.D.

The Jungle Effect: *The Healthiest Diets from Around the World—Why They Work and How to Make Them Work for You*

Farmacology

Total Health from the Ground Up

Daphne Miller, M.D.

wm
WILLIAM MORROW
An Imprint of HarperCollins*Publishers*

Neither the publisher, nor the author, nor any of the medical, health, or wellness practitioners quoted in this book takes responsibility for any possible consequences from any treatment, procedure, exercise, dietary modification, action, or application of medication or preparation by any person reading or following the information in this book. The publication of this book does not constitute the practice of medicine, and this book does not attempt to replace your physician or your pharmacist. Before undertaking any course of treatment, the author and the publisher advise you to consult with your physician or health practitioner regarding any prescription drugs, vitamins, minerals, and food supplements or other treatments and therapies that would be beneficial for your particular health problems and the dosages that would be best for you.

All of the stories featured in the following chapters are based on actual patients in the author's medical practice. However, names and many other details have been changed to preserve their anonymity.

HarperCollins books may be purchased for educational, business, or sales promotional use. For information please e-mail the Special Markets Department at SPsales@harpercollins.com.

A hardcover edition of this book was published in 2013 by William Morrow, an imprint of HarperCollins Publishers.

FIRST WILLIAM MORROW PAPERBACK EDITION PUBLISHED 2014.

Designed by Lisa Stokes
Illustrations by the author

The Library of Congress has cataloged the hardcover edition as follows:

Miller, Daphne, 1965–
Farmacology : what innovative family farming can teach us about health and healing / Daphne Miller, M.D.—First edition.
p. cm.
Includes bibliographical references and index.
ISBN 978-0-06-210314-7
1. Health. 2. Natural foods. 3. Organic farming—Philosophy. 4. Naturopathy. I. Title. II. Title: Pharmacology.

RA776.5.M499 2013
613.2—dc23

2012049844

ISBN 978-0-06-210315-4 (pbk.)

15 16 17 18 OV/RRD 10 9 8 7 6 5 4

This book is written to persons of imagination
and dedicated to Ross, Arlen, and Emet

I'm neither an optimist nor a pessimist. I am a dyed-in-the-wool possibilist! By this, I mean with an eco-mind, we see that everything's connected and change is the only constant.

—Frances Moore Lappé, "How to Think Like an Ecosystem,"
Yes Magazine, April 10, 2012

Contents

PORT ROYAL, Kentucky, is a town so tiny that even the GPS in my rental car seemed unaware of its existence, since it kept offering me alternative locations with "port" or "royal" somewhere in their name. I quickly gave up on technology and dug around in the glove compartment for an old-school, paper map. And there it was, a tiny dot not far from the Indiana border, about sixty miles south of the Cincinnati airport. Leaving the rental lot, I headed along the interstate, passing Big Bone Lick State Park, the Kentucky Speedway, and countless Days Inns. It was not yet 1:30 P.M. local time, but my day had started ten hours earlier and three time zones away in Berkeley. I couldn't help thinking that this was a pretty elaborate pilgrimage for what was likely to be a short conversation. Truth be told, I wasn't even sure what we'd talk about, but even so, I was excited and pushed down on the pedal a little harder.

I found my exit and turned onto a shady two-laner that snaked along, perfectly in sync with its neighbor, the Kentucky River. Although summer was almost over, the rolling pastures on either side remained an emerald green—a jarring sight after the muted patchwork of tans that I'd just flown over in Northern California. I spotted a couple of sagging, tin-roofed barns that I guessed must be defunct tobacco drying sheds and then, about a mile later, two churches, a farm supply store, and a cluster of weathered wooden houses, their wide front porches furnished with rocking chairs. Was this Port Royal? I couldn't be certain, because there was no sign. I bumped over train tracks, took a few more bends, and finally pulled over near a vegetable patch and next to the mailbox of Lane's Landing Farm.

Several months earlier I'd written a letter on cream-colored paper that surely must have landed in this very same box. I'd just finished reading *The Unsettling of America*, a book by writer, activist, and sixth-generation Kentucky farmer Wendell Berry. I left one particular chapter, "The Body and the Earth," so riddled with notes and stars that I could barely read the original text. There were two sentences that I found especially inspiring:

> *While we live our bodies are moving particles of the earth, joined inextricably both to the soil and to the bodies of other living creatures. It is hardly surprising, then, that there should be some profound resemblances between our treatment of our bodies and our treatment of the earth.*

This was exactly what I was looking for! A farmer who had given considerable thought to the links between agriculture and human health. I wrote to Mr. Berry immediately, introducing myself and asking if I might pay him a visit. To my delight, a week later he called me, an overture that I now understand to have been singularly generous, given that he handwrites everything and avoids even the most basic technologies. He explained that he receives many requests and has to be selective about visitors, but a doctor interested in having a conversation about medicine and farming could not be denied. So we set a date, and Wendell Berry told me to come at 2:00 P.M., after midday chores. I should have left it at that, but so great was my excitement that I offered to come earlier and help out. His first response was silence, and then he said, in the politest of Kentucky drawls:

"Ma'am."

No more was needed, as the message was clear: He would not dream of offering to help me do my job as a doctor. Why in the world should I presume that I could help him be a farmer? On my end, I winced with embarrassment and assured him I would not arrive a minute too early. Now here I was, at 2:01 P.M., climbing the steps to the shaded veranda.

Just then, Wendell Berry emerged from behind the screen door. From where I stood, he appeared to be a very tall man, and he was holding a copy of my first book.

"Dr. Miller?" he asked. And silly as I sounded, I could not help but answer, "Yes, and Wendell Berry, I presume?" We both laughed.

"You doctors who kicked over the traces interest me a lot," he said as he settled into one of the white rattan rockers on the porch and invited me to occupy the other. "There are a lot of doctors who are suffering pretty badly, and it is because of that collision of technology with flesh."

In that instant I understood that Wendell Berry knew, better than I, why I was there.

Introduction
Kicking Over the Traces

AS WENDELL Berry gently reminded me, I am not a farmer.

In fact, when I visited Lane's Landing Farm, I would have been hard-pressed to tell you the difference between the humus in the soil and the hummus that goes with pita bread. I certainly could not have defined the term "tilth." I had no idea that chickens prefer to drink cold water, that a dibbler and a squeeze chute are pieces of farm equipment, or that carrots do best when they're grown after beans. But things have changed. I now understand that learning from farmers and experiencing agrarian life can make me a better doctor. I've also discovered how farming, at its best, can offer a bounty of valuable secrets for transforming our personal health and the practice of medicine.

If you can't see the connection, don't worry—you're not alone. My medical colleagues wonder why I've substituted farm time for the more traditional continuing medical education conferences. And when I tell friends and patients that I've been writing a book exploring the links between farming and medicine, their typical response is a polite, yet confused, nod.

Then they say, "Oh, I get it," and bring up their favorite farm-related health concern, such as the connection between antibiotic use in animals and drug-resistant infections in humans, the unknown effects of eating GMO (genetically modified organism) foods, or the contamination of drinking water by fertilizers and pesticides.

While this book addresses all these concerns, no single one is its main focus. Rather it is an exploration of how the art and science of agriculture—from choice of seed to soil management—can *positively* impact your personal health.

Travel with me across the United States, from a biodynamic vegetable farm in Washington State, to cattle and poultry ranches in the Midwest, to an urban farm in the Bronx, and you will discover how healthy soil can produce a healthy immune system; how microbes on the farm can communicate with your resident microbes—or microbiome; how certain grazing practices can produce food that enhances mood and counteracts stress; how urban farming delivers a range of unexpected benefits to boost the resilience of the surrounding community.

On another level, the stories in this book reveal how farming, at its best, can provide radical new paradigms for health and healing. For example, the integrated pest management (IPM) system used by a winery in Sonoma offers a more ecological way to understand and treat cancer. Similarly, the priorities of a pasture-based egg farmer highlight the difference between good stress and bad stress, while his productivity metrics offer a more holistic way to assess well-being. Throughout you will hear the perspectives of researchers from a variety of disciplines, including agronomists, botanists, ecologists, geneticists, immunologists, microbiologists, and nutritionists, who help explain the science underlying these intriguing farm-to-body connections. By the end of this voyage I hope you will have a completely new approach to health and healing, one that starts from the ground up.

MY GERMINATION

So what inspired me to "kick over the traces" and start seeking insight from farmers? Perhaps this is best answered by a brief history of my long-standing (albeit intermittent) connection to the agricultural profession.

First, there was my own germination, which took place on a farm—or more accurately, in a worker's shack somewhere between fruiting persimmon trees, a henhouse, and a braying donkey named Moshe. It was 1965, and my parents had left their hometown near Boston to join the excitement of the kibbutz cooperative farming movement in Israel. There they spent their days picking fruit and tending animals, a time they both distinctly remember as rosy and carefree. They left the kibbutz shortly before I was born. ("We didn't want a communal baby" was my mother's explanation.) But I like to think that this prenatal exposure to an agrarian lifestyle made some kind of enduring impression.

By 1968 my parents had returned to the United States and were enrolled as graduate students at the University of Michigan. One weekend their road trip through the valleys of upstate New York ended with an impulse purchase of a 125-acre farm—the cost of $15,000 ate up my parents' entire nest egg. "The Farm" (as it would thereafter be named) was previously owned by Orson Sheldon, the last in the long line of Sheldons who had homesteaded there since the American Revolution. Sheldon had managed to run the Farm into complete disrepair before cashing in his inheritance. When we took over the land, an ad hoc dump behind the house was strewn with Dinty Moore beef stew cans and liquor bottles, and the early-nineteenth-century farmhouse was so unlivable that it had been converted to a stable—that is, until the horses and cows broke through the rotted oak floorboards and fell into the basement. At the bitter end, both Sheldon and the animals were living in the barn.

Looking back, I can now say that my mother and father were wonderful parents . . . but terrible farmers. On several occasions, spurred on by the

1960s version of our current "back to the land" mania, they tried to homestead on the Farm. They enrolled my brother and me in the local school, planted a garden, and started to collect livestock. But they were urban kids with limited country smarts and few mentors aside from Mr. Lapan, the neighboring dairy farmer who quickly overcame his hippie phobia and would appear in our driveway to offer fatherly advice and shake his head at my parents' mistakes. Needless to say, things on the Farm did not exactly flourish, and our series of homesteading attempts always ended in the same way: As the cold fall days confined us to the farmhouse, the bickering would start, and soon winter would blast in, offering the final dissuasion. By late December my parents would drain the water pipes, sell or slaughter the animals, and head to Washington, DC, in search of yet another post with the Peace Corps. These assignments, although tinged with a sense of failure, were a welcome relief from the snowdrifts and the yoke of daily farming. They also took us to new and distant lands like Morocco, Afghanistan, and Tunisia.

Throughout these exotic travels, and even when we settled in Washington, DC, for my high school years, the Farm remained our home base, the place to which we returned. There my brother and I were free to run amok, and it was most certainly there that my primordial medical brain recorded its first health lessons. Food from the Farm was always connected to a feeling of wellness and renewal: my mother's dolmas, made from the wild grape leaves that covered the old icehouse; the snap peas from the garden; the pure grassy sweetness of the fresh milk I collected by dipping our stainless steel can into the depths of the Lapans' cooling vat. I marveled at the life force I felt from the animals—an electricity that made my skin tingle when I offered a sucking thumb to an unsteady newborn calf in the Lapans' birthing shed or when I stood so close to a cow that her sweet breath misted my face.

It was also on the Farm that I first felt compassion (and sometimes a morbid fascination) for the sick, the suffering, and the dead. One morning

I discovered our only lamb beheaded by a marauding wild dog; I sat in her enclosure, my own head buried deep in my father's arms, and howled in grief. Yet in the midst of my mourning, I clearly recall freeing one eye to check out the network of arteries, sinew, and spinal cord that trailed off into the dirt.

Over the years, distracted by city life and my studies, I spent much less time on the Farm, and by the time I was in my medical training my parents had decided to sell the property and spend part of the year in California. At that point I rarely ventured beyond dimly lit lecture halls or the antiseptic rooms of Boston's hospitals, and my only connection to farming was through my friend and medical school classmate Grant Colfax.

Grant and his three brothers grew up on Shining Moon Goat Ranch in Mendocino County, California, where they were home-schooled by their ex-professor parents, Micki and David. Because of his goats, Grant seemed to know more about medicine than the rest of us. Sitting side by side at the back of the lecture hall, he would fill me in on the goat equivalent of whatever disease was being presented that day. When it was a rare genetic disorder, Grant informed me that his French Alpines occasionally were born with a similar defect. And when we learned about AIDS, Grant was already an expert because his goats were susceptible to the caprine equivalent, a lente virus called CAE. (One can only wonder whether it was in part his early familiarity with the disease that led him to become the director of AIDS prevention for the city of San Francisco and later the AIDS policy adviser to the White House.)

Early on in my medical training I visited Grant at Shining Moon, and in retrospect that trip introduced me to a new way of doctoring. At Harvard Medical School I was taught by a long list of eminent scientists, each recognized for a landmark discovery, but rarely were they role models for engaging with people and keeping them healthy. During an obstetrics rotation, for example, I almost never saw my physician preceptors counsel their patients about diet, exercise, stress management, or other wellness issues. Nor did I watch them support the women in labor, a task that was left to the nurses. Instead, their job was to perform ultrasounds throughout pregnancy and, during the birth process, to interpret data from the monitors in the hall, and to step into the delivery room to catch the baby or—disconcertingly often—to inform parents it was time for a cesarean section. By contrast, on the ranch I observed Grant and Micki discussing their herd and strategizing about how to keep each goat healthy throughout the life cycle. My visit coincided with kidding season, and I was amazed at how Grant rarely strayed from the laboring doe, choosing to spend his nights huddled on a cot in the open-air barn. As I supplied him with sandwiches and coffee, I watched him wince sympathetically through each uterine contraction and then fuss over the postpartum animals and their offspring. Grant, my farmer classmate, gave me a new definition of what it meant to be a caregiver.

Several years later, during my family medicine internship in a community hospital in Salinas—the so-called lettuce basket of California—it was the ugly side of farming that etched itself on my consciousness. While on call in the hospital emergency room, the EMTs brought in on gurneys two young Mexican field workers. The first, a woman, was seizing, her pregnant belly rising and falling like a beach ball in the surf. The other, a man, lay on his side, his body contracted with gut-wrenching pain while a steady stream of drool ran out the corner of his mouth and pooled in a plastic basin. From this macabre scene, the image that most haunts me is their fingers, stained a reddish-black from the juice of strawberry fruit and leaves. This was my first of many encounters with organophosphate pesticide poisoning.

Since then, many of the agricultural practices in the Salinas Valley have changed for the better. But at the time the smell of farm chemicals seemed to be everywhere—smells that I came to associate with bile, seizures, asthma, and a ghoulish assortment of fetal malformations I saw in the hospital's obstetrics ward. Driving home after a long night on call, I would forget where I was and roll down my car window in search of fresh air; suddenly the car would be awash with the odor of diesel fuel and unhealthy earth, soil drenched in ammonia, bromide, and untold other chemicals.

I desperately sought an antidote to these experiences. My husband, Ross, and I were newlyweds, we were both overworked, and our recreation was to create a garden. We cooed over baby eggplants and rejoiced over our first bean sprouts. We suspended a hammock between two pepper trees that bordered our plot and slept there at night, cramped and chilled but lulled to sleep by the sound of new life pushing through the soil. While we took immense pleasure from that garden, our schedules prevented us from mounting a concerted defense against weeds and pests. This lax approach offered its own set of lessons, since I was able to see that nature, even with a minimum of attention, is perfectly capable of reaching her own balance and offering an impressive bounty.

Over the next two decades farming continued to play a role in my life and in my medical career. I've worked with patients to help them find a diet more harmonious with the seasons; I've written about eating traditions in cultures still closely tied to the earth, and I chat each week with my favorite farmers (or at least their representatives) at the Berkeley farmers' market. And yes, I have gardened. But it was one slim book, found serendipitously in the free box outside a local bookstore, that expanded my focus and led me to this story.

THE SOUL OF SOIL

The Soul of Soil, written by Grace Gershuny and Joe Smillie, was such a compelling title that I scooped it out of the free box and, finding a table in a nearby café, I tore through all five chapters in one sitting. It was a guidebook to help farmers and master gardeners improve and care for their soils. What initially intrigued me was the detailed description of a soil ecosystem where the nutrient exchange between soil, microbe, and plant sounded curiously similar to what takes place in our own intestines. Like our own biosystems, it too depends on bacteria and fungi to supply it with the fats, amino acids, and carbohydrates that make up its structures. This was also the first time I understood that the chemical makeup of soil has roughly the same ratio of nitrogen-to-carbon and a similar range for normal pH (6.0 to 7.5) as the human body. Then, midway through the book, it dawned on me that the carbon, nitrogen, and every mineral and vitamin that is a building block in our own bodies is derived from soil. In other words, *we are of the soil.* By the last chapter, I was convinced that this book was not just a farmers' manual but one of the most engrossing medical texts I had ever read. Here was a new vision for how to rejuvenate, rebalance, and heal a complex living organism. I wanted to learn more about how its principles could be applied to me, to my patients, to all of us.

BEYOND DIAGNOSE AND CONQUER

Throughout my career I have sought better ways to approach health and healing. The reductionist medical training that doctors receive is useful when one discrete issue trumps all others—an ingrown toenail, a urinary tract infection, or an appendicitis. Focusing on a single factor—the errant spicule of nail, the bug in the urine, the inflamed pouch of intestine—usually solves the problem. This "divide and conquer" (or "diagnose and conquer") strategy has dominated scientific inquiry for centuries. It is best captured by the seventeenth-century philosopher René Descartes's famous declaration: "Divide each difficulty into as many parts as is feasible and necessary to resolve it."

But most of the time our health needs are more complex and dynamic, just like the soil, and most of what ails us today—depression, anxiety, diabetes, heart disease, fatigue—is multifactorial, chronic, and not well served by a static and highly focused approach. On one level there is our physical makeup: our DNA and the hormones, nerves, and other tissues that manifest this coded information. These structures and chemicals sometimes have an imbalance that can be measured and corrected—such as blood sugars, blood pressures, or hormone levels. But addressing these discrete issues, while important, is only one piece of the puzzle and rarely translates into optimum health. Other important factors include our emotions and mood, whether or not we experience pain, our energy levels, the quality of our sleep, the food we eat, our relationships, if and how often we exercise, the place we live, and the air we breathe.

So how do we put together all these pieces of the puzzle to best serve our health needs? This question has led me to stray beyond the confines of "business as usual" medicine and to try different things over the years. I've restructured my practice to spend more time with each patient (the standard ten- to fifteen-minute medical appointment stems from a widely accepted idea within my profession that only one, or maybe two, issues should be succinctly addressed at each visit), I carefully consider each prescription, and I collaborate closely with medical specialists to streamline my patients' medication lists and help them avoid unnecessary procedures. I've referred patients to nonphysician colleagues—skilled physical therapists, mental health therapists, acupuncturists, nutritionists, osteopaths, naturopaths, and herbalists—who I felt could support their healing process with fewer side effects. And I've attended a variety of holistic medical conferences and workshops, with the hope of finding new models that better address our complex health needs. These programs, while offering me valuable lessons on how to use gentler, nonpharmacologic modalities—such as biofeedback, nutrition, and herbs—rarely provided a new prism through which I could contemplate wellness. Like me, most of my teachers were having trouble breaking away from the reductionist model. I even began to investigate other medical systems—Chinese, Native American, Ayurvedic, homeopathic—that seemed to have a more dynamic and interactive way of understanding health, but I quickly realized that to correctly practice these forms of medicine would require years of study. Moreover, there was much about my own formal training that I valued. What I needed was a new worldview that was still rooted in biomedical science—but I had no idea where to look. Or at least I thought I didn't . . . until that soil guide for master gardeners reminded me about my old friend, farming.

Perhaps, learning from farmers, I could find a better way to maintain balance and wellness within a living system.

AGRICULTURE AND MEDICINE: A COMMON HISTORY

The Soul of Soil set me on a new path of discovery. I started with Sir Albert Howard's classic book, *The Soil and Health,* first published in 1947. Here Sir Albert, who is considered by many to be the grandfather of modern organic farming, shares his observations from decades as an agricultural consultant in India and the United Kingdom. He writes: "The first duty of the agriculturist must always be to understand that he is a part of Nature." Accordingly, Sir Albert lays out these principles: When Nature farms, she recycles everything, never wastes, always leaves a reserve, always has a period of fallow, and always includes animals. He makes the point that, for thousands of years, the most successful farmers have been men and women who were keen observers of Nature and respected her principles.

As I made my way through *The Soil and Health* and other books and articles written by thought leaders in sustainable agriculture, I realized that the professions of farming and medicine grew out of a shared goal: to sustain individuals and communities by supporting the workings of nature and intervening—oh so judiciously—in the cycle of birth, growth, death, and decay. Of course, prior to the scientific revolution the possibilities for "intervention" were fairly limited. For a farmer, they included seed saving, harvesting, cultivating, sowing, herding, and composting, while doctors and other healers might offer solace, assist at a birth or a death, or treat an ailment with a prayer, a behavioral prescription, a diet, or a poultice of herbs.

All of this began to change during the European Renaissance with the advent of modern agriculture and medicine. No longer satisfied with relying on folklore, intuition, and experience, pioneers in both fields (who were often one and the same) developed a scientific method to test observations and to better understand the inner workings of all living matter. From those observations grew the idea that nature, when compartmentalized and scrutinized, will reveal all its mechanisms.

Over the following centuries, this reductionist approach led to great insights in all the sciences—physics, chemistry, mathematics, biology—further contributing to the advancement of medicine and agriculture. In the early to mid-twentieth century, spurred on by the technological demands of two world wars, we saw our most celebrated breakthroughs in both professions. Tanks became tractors; nerve gas became chemotherapies and pesticides; explosives became fertilizers; and powerful antibiotics and antiseptics made an easy transition from war to peacetime use. This was the golden age of reductionist science: These innovations boosted farm yields, alleviating hunger and poverty in many parts of the world, and they offered new medicines to prevent and treat deadly diseases.

But our unilateral focus on solving health and farming problems by subdividing them into smaller and smaller parts has reached a point of diminishing returns. While breakthroughs such as gene mapping, computational science, and nuclear imaging have given us volumes of new information about the inner workings of our bodies and the natural world, they have also generated a greater demand for sophisticated interventions and increased the need for specialists to manage these interventions. This, in turn, has driven costs to an unsustainable level and fragmented important healing relationships, including the age-old connections between farmer and eater, and between patient and healer. Even more perturbing, many of the technologies intended to save lives are now contributing to our modern health woes. This includes an epidemic of obesity, diabetes, and heart disease linked to an abundance of corn, soy, and wheat produced on industrial farms; widespread bacterial, viral, and fungal resistance from antibiotic and pesticide overuse; a nutrient-depleted food supply from overtreated soil; and an explosion of cancers, lung disease, and other chronic ailments associated with the chemical by-products of Big Farm and Big Pharma.

SIGNS OF CHANGE

In response to all of these worrisome trends, things are starting to change . . . at least for farming. I now understand that the whole-system approach that captivated me in *The Soul of Soil* is part of a much greater paradigm shift within the agricultural profession. The enormity of this shift can be measured by understanding that since 2002 California, Indiana, Kansas, Wisconsin, and other agricultural states have seen a 500 percent increase in small to midsize farms that are ecologically or holistically managed. Today, sales of organic produce represent the fastest-growing sector within agriculture. Although the factory farm model continues to dominate, many large agriculture schools, such as the University of Washington and the University of California, Davis, have begun to develop teaching and research programs in sustainable agriculture, focusing on key areas such as water, energy, and soil conservation and integrated pest management (more on this in the following chapters). Even within the political sphere, advocates of sustainable farming have increased their clout. As I write, dozens of organizations representing consumers and farmers are pressuring lawmakers to pass a ratified Farm Bill that ends subsidies to industrial farms while diverting funds to support conservational stewardship and agriculture.

This paradigm shift goes by many names, including holistic, integrated, ecological, and sustainable, but what all these labels share is the idea that a farm is not just a collection of parts but a complex (and sometimes unwieldy) living system. Implicit in this approach is the idea that true health gains for humans and other members of the farm system are to be had by better understanding the interconnections among these living parts and by using the technologies of modern science to subtly enhance them.

Medicine, by contrast, remains largely devoted to the principles of reductionism. Not that I don't see signs of change: The current MCAT (medical college admission test), for example, has expanded beyond a narrow focus on basic science to include more questions on the social sciences, cross-cultural studies, and critical analysis. Recent medical articles have called also for a "complexity approach"—that is, one in which many factors, including biological markers, patient and physician expectations, lifestyle choices, and physical environment, play a role in medical care. Perhaps the most significant effort to shift medicine's worldview has been undertaken by Andrew Weil and his colleagues at the University of Arizona. They have established a health professional training program in integrative medicine that offers courses in nutrition, herbal medicine, and other nonpharmaceutical treatments while exploring new paradigms for health care. Despite these shifts, it is fair to say that most of us in the medical profession are just starting to grapple with what it means to take a "whole system" approach to health and healing. Agriculture, meanwhile, has been considering this question for decades.

From my perspective, there are many reasons why agriculture is ahead of medicine when it comes to weblike thinking, starting with the simple fact that farming, even at its most technological, has never completely turned its back on nature. Growing produce or raising animals on any significant scale must, by and large, take place outdoors, where there is always the possibility of a thunderstorm, a drought, or an infestation. In medicine, on the other hand, we have figured out how to sustain life on a heart-lung machine or in an incubator, and this capability has led us to believe that nature can readily be removed from the equation.

It is also easier to perform experiments in agriculture than in medicine, and farmers can be more nimble than doctors in testing new models and implementing their outcomes. In agriculture, one can make an observation, ask a question, design a study, and potentially develop a radically new approach within the span of one or two growing cycles. In medicine, it can take decades or even generations for the success of a new intervention to become apparent. Ethical, legal, financial, and administrative considerations also make it hard to question or change standard medical practices—even when those practices have unclear benefits and carry considerable risk. (Consider the fact that between 15 and 20 percent of cesarean sections performed in the average U.S. hospital are deemed medically unnecessary or that PSA screening tests for prostate cancer continue to be promoted in many medical centers, despite the lack of evidence that this test either saves lives or alleviates suffering.)

Finally, there is the culture of each profession. Agriculture calls for a certain resourcefulness, the ability to do more with less—a natural result of working within a relatively low-paying and (with the exception of factory farms) meagerly subsidized profession. As a consequence, farming has more than its share of mavericks and go-it-alone cowboys, folks willing to tinker so as to better meet the demands of families, communities, and the soil. In medicine, on the other hand, we frown on those who try to buck the system. One might even say that our guiding oath to "first do no harm," which appropriately reminds us of the sanctity of human life, has been translated into a culture of caution and conservatism.

I must confess that I am not immune to this culture. But once I understood how much the cowboy eco-farmers had to teach me about health and healing, I gained some of their boldness. I decided to take the leap—to put on my rubber gardening boots and return to the farm.

But where to begin? After reading the essay "The Body and the Earth," the answer became clear. I needed to start with the man whom many innovative agriculturists pointed to as their inspiration. I needed to travel to Kentucky to speak with Wendell Berry.

BEGINNING THE CONVERSATION

Wendell and I talked for several hours, sitting in those chairs on the veranda and watching the river pass below and the sun cross the sky. He told me about the foods that previous generations of Kentuckians used to eat ("the family cow, the poultry flock, the garden, and the meat hog—they all fed us"), about his career path from the farm to academia and then back to the farm, and about his three greatest passions: Tanya, his wife of fifty-one years; responsible forestry; and permanent pastures.

Finally our conversation turned to my project. I started by asking him about those sentences that had so inspired me in his essay. Specifically, what did he mean when he wrote that "there should be some profound resemblances between our treatment of our bodies and our treatment of the earth"? Wendell sat and rocked for a while, giving my question considerable thought.

"The critical question that you need to ask in both health care and farming is: What is the pattern you are making? Is the pattern going to be that of a factory, or that of a forest or the native prairie? We go to a place and we say to it, 'Look, you are going to grow corn and you are going to grow it year after year.' We are drawing on natural capital that we cannot replace. We have done the same thing with our health. We stop listening to ourselves and submit to that equation: organism equals machine. What happens then is you get some doctor who will cut your guts out of you without even looking at you."

He pointed off into the distance, to a rolling pasture dotted with sheep. "Take this hillside here. It is always in pasture now. It was row-cropped in earlier times, but that was a mistake."

Truth be told, this was the first time I had thought about "factory medicine" as the counterpart to "factory farming." But the concept made perfect sense and captured much of what I viewed as the shortcomings of our predominant model for health care: the wasting of resources, the overemphasis on pharmaceuticals for short-term benefit at the risk of long-term side effects, the focus on organs rather than organisms, and a general disregard for the body's natural ability to heal. I told Wendell Berry that I wanted to learn from farmers who were more holistic and patterned their farming on nature. I hoped to apply these lessons to my medical practice.

He nodded.

"You know, those terms 'organic' or 'holistic' you can wear on a T-shirt. But what people are trying to mean by those words is a kind of courtesy or respect, always accompanied by affection. When we make a decent marriage or friendship or farm, we make a partnership. In such a partnership, we look and listen to what our partner is trying to say back to us. We have a conversation where we are working for a mutual benefit."

Stressing the word "mutual," he leaned down and petted the border collie at his feet.

Just then, Tanya, his wife, appeared on the porch, having returned from town. In her presence, I watched Wendell transform from a farmer-philosopher to a dancing-eyed, droopy-grinned schoolboy. They talked about her day, which was largely spent ferrying grandkids, and then, with that silent accord that inevitably develops after fifty-one years of a good marriage, they both headed down the steps to start their early evening chores.

Happily I trailed behind, trying my best not to meddle as they checked on their sheep, fed the pair of llamas charged with protecting their flock, and tied up some loose boards in the barn with a piece of bailing twine.

"Daphne, there is very little in life that cannot be fixed with a good length of twine," Wendell said as he gave a good final yank on his square knot.

Then he and I headed out to the pastures to move his two massive Percheron horses, animals he uses to pull his plows. Wendell's long legs easily cleared the same barbed-wire fences that I had to shimmy under. On our way, he bent forward and pulled up a clump of pasture, proudly showing me the old symbiotic patches of bluegrass and white clover.

"These have been slowly reappearing ever since we put this overtilled land in permanent pasture," he explained. "Land that is under perennial cover is safe. The sod thatches like a rug on top of it and protects it from erosion. Perennials are deeper-rooted than annuals. They bring up nutrients and moisture that annuals can't reach."

When we returned to the house, Tanya was washing greens from the garden and beginning to prepare supper. In the warm evening light of the kitchen, I thanked them both, we hugged good-bye, and they generously invited me to come back when I had finished researching this book.

Then, with his hand on the screen door, Wendell gave me one last thought. This was his parting gift to me, a series of questions that I could rely on, as I journeyed from farm to farm, to start the conversation with farmers:

"You always need to 'consult the genius of the place,' " he said, quoting the poet Alexander Pope. "You should ask the farmers, 'What was here when you came? What was here before you came? What was here for you to start with? What does nature require of you here? What will nature help you to do here?' "

I'd read other interviews and essays in which Wendell Berry used this quotation from Pope, but in the context of what I was undertaking it had a whole new meaning. I climbed down the steep steps toward the river and my car, nearly falling on my face as I scribbled his last words into my notebook.

1 Jubilee

CARNATION, WASHINGTON

What a Biodynamic Farmer Taught Me About Rejuvenation

It is now becoming clear that taking [calcium] in one or two daily boluses is not natural, in that it does not reproduce the same metabolic effects as calcium in food. The evidence is also becoming steadily stronger that it is not safe, nor is it particularly effective.

—Ian R. Reid and Mark J. Bolland, *Heart* journal

I ARRIVED in Seattle, en route to Jubilee Biodynamic Farm, on the last dreary day of a strange midsummer cold spell. Erick Haakenson, Jubilee's owner and head farmer, greeted me on the curb of the SeaTac Airport with a bear hug, apologizing for the weather and for the fact that he was still hungover from neighborhood festivities the night before. Until that moment, our only connection had been a couple of e-mails and one fuzzy cell-phone conversation during which we discussed the details of my upcoming two-week internship. When I asked him how I would recognize him at the airport, he said, "Ah, just look for the typical Norwegian farmer."

And now there he was, tall and broad with muddy boots, bottle-blue eyes, and straight yellow hair, a color that was especially striking given his white beard and sixty-two years. (Later on I couldn't resist asking him if it was dyed, a question that rightfully earned guffaws from all farm interns within earshot.) It was all as I had imagined, the only disappointment being his vehicle, a fairly clean VW wagon. Where was the requisite dusty pickup, its bed overflowing with hay bales? No matter. I quickly got over this minor letdown, telling myself that I'd get my fill of farm equipment in the weeks to come. I loaded my backpack and my own too-clean work boots into the hatch, and we headed east toward Jubilee.

Like many of the farmers I would meet, Erick had taken a circuitous path to farming. As we drove he told me about growing up in Linwood and Edmonds, Washington, going to the University of Notre Dame to study philosophy and the Great Books, then switching paths to become a professional fisherman, hauling salmon out of Alaska's freezing waters. From there he veered back toward academia, landing in the religious philosophy program at Yale Divinity School. But the life of a scholar felt too confining after all those years at sea, so finally, in 1989, he decided to fulfill a lifelong dream and become a farmer. That's when Erick bought Jubilee and moved onto the land with his four teenagers and his first wife, who from the start did not share his agrarian aspirations.

As we left the highway and started down a narrow country road, Erick told me that his first years at Jubilee tested his mental and physical limits far more than his time at sea or the semesters he spent in high-powered academia. His marriage fell apart and he struggled to turn his twelve acres into a viable enterprise. But everything got better after Wendy, lovely and four years divorced, magically appeared as a farm intern. From what I could gather, Erick fell for her immediately. Wendy, he said, was a natural farmer, and she came with the bonus of other family members, including her mother and two daughters, who lived nearby and often helped with the biweekly farm stand. Erick's daughter also lived on the property with her boyfriend, who was the main farm supervisor. His description gave me the impression that they happily coexisted, like some rural adaptation of The Brady Bunch.

Within forty-five minutes, we pulled up in front of a newly built farmhouse on stilts. It was a perfect square with windows on all sides and a wraparound porch that looked out to the fields. (Stilts, I soon learned, were there to accommodate the winter floods that routinely cover the entire valley.) Wendy, on a midmorning break, greeted me in the mudroom, where she was pulling off her soggy jacket and soil-caked rubber boots. With her hair in a tight bun, she reminded me of the comely, apple-cheeked farm ladies that decorate tins of Danish butter cookies.

The three of us went inside and made strong coffee, then sat in the cozy overstuffed chairs in their living room, blowing on our mugs. From our very first correspondence, Erick didn't seem to think it odd that a family doctor should want to learn from a farmer. But Wendy was curious. What exactly was I doing there? What did I expect to get out of the experience? In the face of these direct questions, I had no choice but to rely on Wendell Berry:

"I guess I'm here to consult the genius of this place," I told them.

I gazed out through the drizzle to the expanse of glistening fields with a peek of the river beyond. They both seemed satisfied with this answer.

GOOD TILTH

The following morning I awoke in a patch of sunshine; summer had finally returned to the valley. I found Erick near the barn, assigning the day's chores to the Jubilee interns. He walked me over to the nearby toolshed and introduced me to the Dutch "hula" hoe, an odd-looking tool with a blade shaped like a stirrup. Then he sent me out to weed the broccoli and other brassicas alongside a dozen or so of the "work-shares"—people from the surrounding community who traded labor for a weekly basket of fresh-picked produce. In my crew was a dog trainer, an acupuncturist, and a software engineer, all united by their love of farming. Soon I was talking with a tanned man wearing a red bandana who was hoeing directly across from me. John Romanelli described himself as a woodworker with the heart of a farmer, and he told me that he was so passionate about good soil that he had named his company Timbers and Tilth.

"What is tilth?" I asked John, taking care not to injure the broccoli as I used my hoe to delicately clear away some weeds. He stopped his work and squinted at me, wiping his brow with his bandana. I blushed, having outed myself as a farming ignoramus. But I was being too sensitive because John didn't seem to think it was a stupid question. He leaned on his hoe handle and gazed thoughtfully off into the distance. Finally, he knelt down on one knee, scooped up some earth in both hands, and passed it to me across the row as if he were making a religious offering. It was a deep cocoa color, and each particle squiggled and bounced, clearly powered by micro-creatures hidden within. Despite the hot sun, it was moist, with a texture not at all like the uniformly minced, lifeless particles that make up a store-bought potting soil. It reminded me of a well-made pastry dough before it's been kneaded or rolled, with lots of odd-shaped clumps of flour and butter, separated by plenty of air. I noticed that my mouth was watering, and I laughed. What an odd reaction to have to a handful of dirt!

"This is good tilth," said John. "It's kind of like good Chi or good health. You have a hard time describing it, but you know it when you see it. You can look at your patients and tell when they have it. Well, it's the same for us as farmers."

FAT OLD MAN LAND

It turns out that Jubilee had not always had such excellent tilth. The following day, after finishing early morning chores, Erick and I sat down at the dining room table to enjoy our plates of scrambled eggs, freshly plucked from the "chicken tractor"—the farm's itinerant henhouse. Erick looked more disheveled than usual. He told me he'd slept poorly because he had spent the night worrying that his cows would escape their enclosure and destroy his neighbor's fledgling vegetable farm. As part of his biodynamic system (more on this later), he moved the animals to a fresh pasture daily, but the most recent pen was dangerously close to the property next door. With ten minutes of freedom, his cows could trample and devour an entire season's profit. What made Erick's sleep even more fitful was the knowledge that this couple already teetered on the brink of bankruptcy.

Erick's other neighbor, Van, the one who hosted the big party the night before my arrival, drove up in his pickup and joined us at the table. Van has farmed in Snoqualmie Valley for more than forty years and is considered the mayor of the road. He's also a tractor mechanic, a highly prized skill in any farming community. That morning he wanted to powwow with Erick about how to halt a local dam construction by the U.S. Army Corps of Engineers, a project they both believed was responsible for the previous winter's record flooding on the valley floor. After the two farmers finished their conversation, Van started to tell me about Jubilee in the early days.

"When Erick took over this land, this soil was a mess. There were old tires and chunks of concrete, dirt bike tracks, berry bushes and thistles. It was like starting with an old fat man and trying to get him into shape." Van sighed, spreading his arms over his own generous belly.

Erick seemed to appreciate the metaphor. He chuckled and explained that when he purchased his twelve acres, he had not wanted to leave anything to chance.

"I was so darn worried about this soil. I knew it was depleted, and I asked myself: 'Do I want to be a part of the crowd that's growing food that lacks nutritional punch?' I knew that nutrient-dense food could not come from nutrient-deficient soil."

TO VAN'S HOUSE & TO

SNOQUALM

COWS

JUBILEE FARM

LOAFING SHED

PIG PEN

PARK HERE FOR FARM BOX PICK-UP

MATT & DEANNA'S FIELDS

MANURE

HERBS

PEAS

BORAGE etc.

ALBION

WORM TUNNEL & CASTINGS

GOOD TILTH

OM

ROOT HAIRS

BACTERIA & FUNGI

Jubilee Farm

Trading in Nietzsche and Kant for tomes on agronomy and pest management, Erick threw himself into learning how to farm with the same intensity he'd focused on his graduate work in philosophy. He was especially impressed by the writings of soil pioneer William Albrecht, who proposed that there is a "golden mean" for topsoil—an ideal ratio of vitamins and minerals, whether in Toronto or Timbuktu, and whether one is growing rice or radishes. Eventually Erick's reading led him to a student of Albrecht's, who had taken things a step further and founded a soil analysis firm. His company tests fields and recommends the type and quantity of ammendments needed to achieve a perfect "Albrechtian" balance.

"I invested in a big way in this testing method," said Erick. He disappeared briefly into his home office, then came back with a thick manila folder that he plopped down next to my coffee cup. Glancing at the first sheet in the pile, my immediate thought was, *Why is Erick sharing his medical records with me?*

The form looked almost exactly like the lab readouts I get on my patients. There was a column that listed minerals: calcium, magnesium, potassium, nitrogen, and sodium. There was also a column for results and a third column with ranges for normal. It was only when I looked more closely that I noticed some important, but subtle, differences. The letterhead bore the title "Agricultural Services," not the name of a medical lab, and in the upper left corner, where I was used to seeing the patient's name and date of birth, I found instead the following information:

Location: Jubilee Farms
Crop: Tomatoes
Field Sample 1/G-2-05

Each page in the pile was identical in format, but the crop names, field samples, dates, and of course the test results themselves were different. An inset on the page gave the recommended pounds per acre of nutrient required for that field: "sulfur 85, calcium 2,000," and so on. On the bottom of the sheet was a general directive: "All recommendations are to be soil-applied and broadcast unless otherwise specified."

A JUNKED-UP, WORN-OUT BODY

As I leafed through Erick's reports, I thought about a patient named Allie whom I'd seen for the first time just one week before taking off for Jubilee. She was the physical embodiment of Erick's dirt bike–scarred, tire-strewn land. She looked as if she were in her midfifties, although her chart told me she was only forty. Her dark bangs framed a sad, pale face with gray circles under the eyes, and her voice often dipped to a whisper as she described her chronic bloating, allergy symptoms, weight gain, and fatigue. If I were to sum up her condition with one word, it would be "depleted."

Allie had never experienced a dramatic illness or accident, and she was unable to trace the onset of her symptoms to a specific day or week. Rather, as she explained it, "I probably had all this going on at a low grade for years."

But she had largely ignored her problems. At first, she was distracted by the demands of starting her own specialty textile business, and then, when her father was diagnosed with a terminal disease, she was overwhelmed by heartbreak and perpetual jet lag from her many cross-country flights to visit him. One day she hit the wall. She stayed in bed for two days but felt no better. When she finally dragged herself from under the covers, she headed to her doctor instead of her office. That day she began to undergo what would ultimately add up to more than one hundred tests. Like Erick, she had organized her reports in chronological order in a manila folder, and on her first visit to my office she handed the whole stack over to me.

At the top of the pile were the standard "fatigue" lab tests: chemistry panels, thyroid tests, vitamin B_{12} and D levels, and tests for anemia and autoimmune disease. The results did show that Allie was somewhat deficient in vitamin B_{12}, vitamin D, and iron, and she had been hopeful that correcting these numbers would reinvigorate her. But after several months of high-dose supplementation, she felt little improvement and went on to consult (in sequence) with a gastroenterologist, a rheumatologist, an endocrinologist, a neurologist, an allergist, a naturopath, and a chiropractor. With each provider, the tests became more detailed and more exotic: hair analysis for heavy metals such as mercury and arsenic; saliva levels for hormones; blood tests for herpes virus, Lyme disease, babesiosis, food allergies, trace vitamins and minerals, and various neurotransmitters and amino acids. The gastroenterologist had ordered a hydrogen breath test, an endoscopy, a colonoscopy, and even a blood test for a couple of genetic markers that are linked to bowel problems.

But the results yielded little. The hydrogen breath test suggested Allie might have a small bowel bacterial overgrowth, and so she took two weeks of a prescribed antibiotic and was told to avoid gluten, dairy, and gas-producing foods, including broccoli, Brussels sprouts, leeks, onions, and beans.[1] The in-depth micronutrient analysis reported that for fifteen of the measured items, her blood levels were lower than the lab's perfect norm, so she dutifully took vitamins and minerals to boost these numbers. But any improvement was fleeting and she quickly reverted back to her sick, tired baseline. In fact, her digestion seemed to be getting progressively worse, to the point that she felt as if her intestines could barely process anything. The only foods she could eat without getting belly pains were energy bars, canned chicken soup, and steamed spinach and chicken. She was also going broke. She had already spent several thousand dollars out of pocket on her medical expenses and was too sick to work. She asked me if I thought she needed more tests, or if there might be a different approach.

Recalling this exchange, I continued to leaf through Erick's reports, finally getting to the last sheet in the pile, an analysis from 2004. This was curious. Was there another volume of reports that brought us to the present? Or had Erick at some point abandoned the "test and replace" approach to soil improvement and turned to something else? How, in the end, had he transformed his "fat old man land" to soil that, according to John Romanelli and others, had such excellent tilth? I looked up to ask him, but he and Van had disappeared. Clearly it was time to get to work.

TESTING RECONSIDERED

I headed to the pea patch, where a couple of members of the full-time farm crew were "nurturing" the vines. I immediately knew what to do, thanks to a fine lesson on nurturing I had received the previous day from Casey, a farm assistant from Connecticut with a lumberjack beard. We were standing in the steaming tomato house, sweat rolling down our sides, as I watched his callused fingers gently lift the plant's leader, being careful not to injure the delicate flowers or tendrils, and then deftly snip off the suckers—any wayward growths that try to suck energy from the main "leader" vine. It reminded me of all the times that I, as a green medical student or resident, stood there motionless (and in a similar state of physical discomfort) while a senior physician taught me a new procedure.

Tending to peas requires less concentration and endurance than tomatoes, and my focus wandered as I made my way down the row. I began to make an inventory of all the flora I could see within a ten-foot radius. First I admired the neighboring plots of scallion and strawberry—they, like the peas, seemed to be thriving. Next I took in the marigold and borage that bordered the pea vines, floral decoys intended to distract insects from the nearby vegetables. Then, looking down, I noticed that the rich soil underfoot was not just soil: It was host to hundreds of weeds sprouting willy-nilly. Some were velvety, and others were jagged and prehistoric-looking, with tiny flowers. Just in this one small corner of the farm I could spot dozens of plant species, each one with a slightly different root depth and nutrient requirement, and each one most likely contributing to the good tilth of Jubilee.

I stopped picking and munched on a pea. Could a soil test accurately gauge the needs of all these plants? And how about all the microbes and minute animals that had animated John Romanelli's handful of "good tilth"—did the tests even consider their requirements? Finally, I thought about Allie and the complexity of her body—indeed, the complexity that exists in all of us. Did the sum of her tests add up to a picture of what was wrong with the whole person? The fact that she was no better seemed to suggest otherwise.[2]

CONE SPREADERS AND BAGS OF VITAMINS

That night, in Erick and Wendy's kitchen, I made grilled kale with faro and mint. Matt and Deanna, the young couple whose vegetable patch had triggered Erick's insomnia, joined us for dinner, contributing their first tomatoes of the season and a memorable local goat cheese called No Woman. The folder of tests, still sitting there on the dining table, prompted Erick to continue the conversation where we had left off twelve hours earlier.

"I spent days following up on these computer readouts," he sighed, fanning himself with the stack of papers. The chilly weather that had greeted me at Jubilee had been replaced by 100-degree heat, and even at eight in the evening it was still a little too warm in the farmhouse. "I used my John Deere cone spreader and ended up putting thousands of pounds of minerals on my twelve acres." (He was not exaggerating. I estimated from the reports that in the years Erick used the agricultural testing services he spread more than fifty tons of imported minerals over his land.)

"But somehow it didn't feel right. There were lots of minerals that I wasn't sure where they were from. They were probably taken from developing countries, where their soil needed these minerals more than we do. I was also wondering, *If these are all so good for my plants, why does the manufacturer recommend that I wear a mask when I'm spreading them?*"

Plus, despite all his efforts, Erick was not seeing the miraculous improvements he'd hoped for.

"I couldn't help thinking, *Yeah, I'm putting these minerals on the soil, but are they really getting to the plant?* And if I happened to put down a little too much of one thing, what did it do to all the other nutrients? I'd heard stories about how adding too much of one thing can 'lock up' other elements. This could create soil conditions that were even worse than when I started."

Once again, I thought of Allie. If she were sitting at the table listening to Erick, I am certain she would nod in agreement. To her first appointment in my office she'd brought in not only her thick file of test results but also two shopping bags full of prescription pills and over-the-counter supplements. One by one she unpacked them onto my desk and the neighboring bookshelf, until my little exam room looked more like a Vitamin Shoppe than a doctor's office. The pharmaceuticals I could recognize right away: a proton pump inhibitor for her stomach, an antispasmodic for her lower abdominal cramps, an antidepressant for her mood, and an antihistamine for her allergies. But a good proportion of the bottles in Allie's pharmacy were labeled with vague names like "Vital Force" or "Woman's Thrive" rather than a specific nutrient. And when I read the fine print, I saw that some nutrients reappeared on a multitude of bottles. For example, I found five supplements that contained zinc and four others that listed vitamin A or its metabolite, retinoic acid (a nutrient that can cause bone loss at higher doses).

Looking at this impressive array, it was not much of a stretch to think of Allie as "locked up." Indeed, any drug or supplement can have unintended effects, and humans, like soil, can become deficient in one nutrient as a result of having too much of another. For instance, excess calcium can create zinc and iron deficiencies in humans. (Interestingly enough, Erick told me that excess phosphate in the soil can create the same deficiencies in his plants.) I wondered how many adverse interactions might be taking place in Allie's fragile system because of all the drugs and supplements she was taking.

This is not to say that vitamin supplements have no value. Consider the sailors of yore who, after several months at sea without fruits or vegetables, all developed the weakness and bleeding gums associated with scurvy (vitamin C deficiency). Most certainly, a good supplement would have prevented this problem. Vitamins and enriched foods have been shown to play an important role in improving life expectancy in developing countries where food shortages have led to malnutrition. Similarly, people who have had gastric bypass surgery or who suffer from cancer or chronic gastrointestinal or kidney problems often have severe nutrient deficiencies. For them, supplementation with specific vitamins and minerals is in order. But for Allie and for others who might be suboptimally nourished but are not profoundly vitamin-deficient, using vitamins, minerals, and other supplements to achieve perfect balance has proven to be as disappointing as Erick's attempts to use a "test and replace" system to achieve perfect soil health.

So far, pretty much every large, well-designed, randomized study has shown that supplements do little to improve health or prevent disease—and in some instances may make things worse. B vitamins have failed to lower the risk of heart attack and stroke; quite the opposite, the chance of stroke was higher in study participants who took B vitamins. Vitamin E did not ward off heart attacks and was linked with greater mortality overall; high-dose beta carotene actually boosted the rate of heart attacks and lung cancer, and calcium pills were associated with more plaque, or buildup, in the arteries. (All these findings are especially vexing given the consistent findings from population-based studies that people who eat nutrient-rich foods have low rates of all these diseases.)

If "test and replace" was not the answer, what could help someone like Allie get to a better state of "tilth"? Perhaps she could borrow a lesson from Jubilee. What I needed to know was this: what had Erick and Wendy done to their "fat old man" land to finally return it to such excellent health?

AN EVENING WITH RUDOLF STEINER

Over dessert, I asked them this very question.

We were all sharing a basket of the Albion strawberries that I had picked right at dusk. (I should add that all four farmers teased me about my selection of berries, many of which they deemed unripe. For anyone who lives close to strawberry plants and has the luxury of waiting, "ripe" means almost falling off the vine. But how could I have known this, since I don't grow them in my own garden and the ones I get in my local market are usually picked before they are ready?)

I asked Erick and Wendy whether there had been a game-changer when it came to the health of their land.

"What really made the difference," Erick said, "was moving to BD [biodynamic farming]. We converted to BD somewhere around the time we stopped testing, and that's when things really started to get good. In BD the animals replace the additives, and they work much better."

"Tomorrow you should go spend some time with the cows," he added as he got up from the table and headed off to bed.

Of course, I knew this was a biodynamic farm. In fact, if you go to Jubilee's website, it says "Jubilee Biodynamic Farm, Inc." But the truth is that in the hectic weeks leading up to my visit, I barely had time to Mapquest Jubilee in Washington State and figure out how to get there. I'd utterly failed to educate myself about biodynamics. The name sounded healthy and wholesome, and I assumed that it had something to do with organics, but obviously I needed to learn more.

That night, sprawled on my bed in the farmhouse guest room, I surrounded myself with a collection of books on biodynamics, all pilfered from Erick's impressive library. Rudolf Steiner, the founder of BD, was born in rural Austria in 1861, and his writings and teachings covered a range of topics, from religion to education to medicine to farming. In all these disciplines, Steiner was an oddball who challenged the reductionist worldview held by most of his colleagues. Instead, he proposed a complexity model, urging scientists to explore the many connections between the physical and the spiritual and between humans and other living systems. If one were to diagram Steinerian relationships, they would look like an intricate, three-dimensional spiderweb—a striking contrast to the linear algorithmns that fill most medical texts.

In his "Agricultural Lectures," a series of talks that he gave to German farmers in 1924, just one year before his death, Steiner expanded on his theories and also gave practical suggestions for applying them to the farm. He predicted that the widespread adoption of nitrogen-based chemical fertilizers, an innovation that was greeted by his contemporaries as nothing short of miraculous, would be a shortsighted solution that quickly degraded the health of the soil and, by extension, animal and human health. He explained that to preserve soil fertility and human vitality, each farm needed to be a self-supporting eco-cycle, or self-powered organism. (In Greek, *bio* means "life" and *dunamikos* means "power.") In its ideal form, biodynamic farming has no need for outside additives or inputs such as fertilizers, pesticides, herbicides, or gasoline—the true definition of "sustainable." To accomplish this Steiner offered suggestions well supported by existing agronomy research, such as using compost to drive the nitrogen cycle. But other recommendations veered sharply away from established science. For example, to boost fertility, he told farmers to pack herbs and manure into cow horns and bury them at various stages in the zodiac calendar, then unearth them and spray them in extremely dilute concentrations on their fields. Perhaps it was this kind of talk that caused Steiner to be viewed as a kook by most of his colleagues and by many in mainstream science today. (A Northern California wine producer maintains a blog entitled Biodynamics Is a Hoax in which he refers to Steiner as an "LSD-dropping Timothy Leary with the showmanship of a P. T. Barnum.")

But as I read on, I could not help noticing that a surprising number of Steiner's seemingly bizarre statements have proven, with time, to be accurate. For example, he cautioned that cows fed protein from other animals would go mad. Now we understand that mad cow disease is caused by prions—viruslike particles transmitted to cows when they are given commercial feed fortified with ground-up nervous tissue from other animals. Steiner's predictions about nitrogen fertilizers have turned out to be equally prescient. Today, from India to Indiana, huge swaths of land are unfarmable because fertilizers made from fossil fuels have so destroyed the soil.

I closed my book on biodynamics and turned off the light. But, despite the day's hard labor, sleep did not happen. I tossed and turned in the still-hot air, trying to get comfortable, and just when I started to drift, a nearby pack of boisterous coyotes jolted me upright. I wandered out onto the wraparound porch and sat there in my T-shirt, hoping for a breeze and listening to the round-robin of howls echoing through the valley.

Nothing I had just read was that new or surprising, and then I realized why. Cow horns and the zodiac aside, Steiner's ideas practically mirrored those put forth by Albert Howard in *The Soil and Health*, the book I'd read months earlier when I first became inspired to spend time on farms. Steiner's tenets were also captured in the way Wendell Berry treated his pastures in Port Royal, Kentucky. Sir Albert, who originally published his book in 1947, does not reference Steiner. (It is unclear if he was aware of Steiner's work.) And when I asked Wendell Berry about biodynamics, he told me he knew little about it. But if you look at the life experience of all three men, it makes sense that each would have arrived at a closed-cycle model for healthy farming, since each had spent time observing and learning from traditional farmers. (Steiner had lived in rural Austria; Sir Albert was inspired by what he saw in India and China; and Berry had learned from wise old-timers in the hills of Kentucky.) No matter their location, these peasant farmers practiced a kind of agriculture that relied on examples from nature rather than a technical manual. In each instance, they worked the land without access to pesticides, herbicides, and chemical fertilizers, and so they did what was logical: They recycled their fertility back into the soil.

These farming systems had survived through the generations because they preserved the health of the land, the animals, and, yes, the people.

Without warning, a gentle breeze, laced with ripe strawberry, found me in the darkness. The smell reconnected me with the field beyond the porch, its soil alive and quivering in good tilth.

I decided to take Erick's advice. Tomorrow I would visit the cows.

THE LAW OF THE RETURN, OR THE ETERNAL DANCE OF THE MICROBES

The next morning I tagged along with Ian, the farm assistant, who was charged with moving cows to that day's fresh enclosure. As we pounded in the stakes for the next electrified perimeter, I noticed how the animals lowed and quivered with anticipation for what lay just ahead. We rolled back the wire that separated the old and new grass, and they moved in at a polite jog. After all, what was the rush? They would have all day to sunbathe and chew and chew. From my first day at Jubilee, I understood that the cows were there to help fertilize the fields, but after my evening of study, I more clearly saw their role in the farm's biodynamic cycle.

With each bite, a maceration of greens, saliva, bacteria, and other organisms travels from the field into the great distillery of the cow's rumen. Here, things are further broken down by a great churning motion and by billions of intestinal bugs, or microbiota. (These microbes are so plentiful that their DNA outnumbers the animal's own somatic cells 100 to 1, a fact that is also true of humans.) The microbiota, taking their cues from the mucosal cells that line the cow's intestine, begin to process the plant material and make it useful to both microbe and cow. They package or synthesize essential vitamins, antioxidants, starches, and proteins in a form that is ideally suited for that cow; make a series of proteins and sugars that protect the intestinal wall (and the rest of the body) from infection or allergy; and synthesize enzymes that help regulate the cow's metabolism. In return, the microbes are nourished by digested plant material and by carbohydrates released from the bovine intestinal cells. In short, what goes on in the cow's gut is a shining example of a symbiotic, microbe-host relationship.

After much churning and absorbing, what was once a mouthful of pasture grass hits the ground in two forms: first, as a dousing of urine, which is so filled with antioxidants and antibiotic-like substances that biomedical researchers are actively mining its pharmacologic potential, and then as a perfect "cow pat," a mineral-rich cake. The soil microbiota, along with worms and tiny mammals (collectively referred to by farmers as "soil biology"), consume and decompose the manure to form the soil's rich top layer, or humus.

The biodynamic cycle is complete when the microbiota in the soil coordinate around a plant root system belonging to a new blade of grass or, since Erick eventually converted his grazing pastures to vegetable plots, a seedling of broccoli, beet, or spinach. What is formed is a utopian community called the rhizosphere; here the microbiota dine on substances produced by the growing plant, but like the biota in the cow's intestine, they reciprocate by producing thousands of enzymes and antioxidants that nourish the plant, boost its immunity, and protect it from pests and unfriendly weeds. (I'll have more to say about these immune boosters and natural pesticides in the following chapters.) The microbiota also communicate with the plant's roots to offer them an ideal dose and combination of "chelated" micronutrients. Chelation, from the Greek word for "claw," means that inorganic minerals such as zinc and iron are packaged in the center of a larger carbon-containing molecule. In this form, they are delivered in a safe dose to the plant without causing toxicity or vitamin deficiency or "locking up" other nutrients. Chelation happens naturally in the soil but is hard to replicate effectively in a lab or factory.[3]

All this explains why Erick has traded "test and replace" for cows and fungi and bacteria and nematodes: He has discovered that no amount of chemical additives or mineral supplementation can hope to duplicate what naturally occurs in his well-honed rhizosphere. The best thing he can do is leave his soil alone and allow microbiota to drive the life force on Jubilee.[4] Then, as long as the farm denizens—cow, chicken, or human—continue to eat from this soil and cycle their waste back into it, their systems will keep humming as nicely as the rhizosphere.

Sir Albert, Rudolph Steiner, and the generations of traditional farmers who inspired their work knew all this intuitively, but now Erick has some field research to back up these ideas. During a twenty-one-year experiment, Swiss researchers showed that biodynamic fields had a higher microbial biomass and more nutrient availability in the soil than neighboring conventional fields that used synthetic fertilizers and imported minerals. The biodynamic fields also required 20 to 50 percent less fertilizer and fossil fuel–derived energy input than conventional fields to get the same crop yield. What was most surprising was that the biodynamic field had a healthier soil and richer compost than neighboring plots that were certified organic but had not been cared for using the principles of biodynamics. Other studies in the United States, Italy, and New Zealand have reproduced these results.[5]

WHAT ABOUT ALLIE?

Standing in that field, watching the hale-looking cows, it occurred to me that my patient Allie's subtle vitamin deficiencies, her "wilt," her "bloat," and her allergies might all be linked to disturbances in her intestine and her microbiota. I wondered how the lessons I was learning about eco-cycles could be applied to her health.

Allie was definitely not a part of the Jubilee cycle or any farm cycle. In fact, it would have been difficult to trace any ingredient in her nutritional supplements or her energy bars back to its field of origin. Most of those ingredients probably originated in a large corn or soy plot somewhere in the Midwest or maybe Brazil. (Yes, corn and soy are hidden in an astounding percentage of our foods and are even the building block for many off-the-shelf vitamins.) Most conventionally grown corn and soy available in the United States are exposed to or producing—as is the case with genetically modified plants—generous helpings of pesticides and weed killers, substances that overwhelm the delicate subterranean microbes. Allie's standard "warm" salad and soup, made from organic spinach, carrots, and chicken purchased at Walmart, were also unlikely to connect her to a healthy farm cycle. Almost without exception, the organics offered by large chains are produced on industrial-size farms that do the minimum required to maintain certification (see the list below on organics). Instead of maintaining an eco-cycle, these large-scale growers use a "test and replace" method; instead of intercropping and rotating crops, they tax their land's nutrient reserves by planting acres of the same fruit or vegetable year after year; and instead of using a no-till method, they plow their fields, sending much of the rich top soil into the watershed, where it can do little to nourish the next generation of plants. In short, it was doubtful that the soil producing Allie's food looked anything like that found at Jubilee.

So what if Allie changed the food she ate and joined a farm eco-cycle? Would she derive the same benefits as the cows, the plants, the soil, and all the other players on Erick's farm? What would happen in her gut? How would this affect her health in general? In the weeks that followed my visit to Jubilee, I decided to find out.

What are you guaranteed when you buy "certified" organic food?

- That the farmland used to grow the food was not treated with synthetic fertilizers for three years prior to harvest
- That the crop was not irradiated, grown from genetically engineered seed, or fertilized with sewage sludge
- That organic meat was not raised with hormones or routinely dosed with antibiotics

What are you *not* guaranteed?

- That the food is freshly picked from a local farm
- That the farm produces its own fertility and avoids importing minerals or fertilizers or "natural" pesticides
- That the farm recycles its water and other waste
- That animals were raised outdoors in pastures
- That the farm uses no-plow or no-till soil management
- That the farm has fair hiring practices and protects the safety of its workers
- That prepared foods labeled "organic," like cereals or cookies, have fewer calories or preservatives or are more nutritious than conventionally grown food

Based on the US Organic Foods Production Act of 1990

THE GNOTOBIOTIC FARMER

In the sciences, the loneliest place to venture is the no-man's-land between two contiguous fields of inquiry. The vast trove of data and hypotheses falls away, and one enters an information wasteland. This is exactly where I found myself as I tried to understand what would happen to Allie's health on both a micro level and a macro level if she became a part of an eco-farm cycle such as Jubilee's. There was plenty of information about soil ecosystems and about human ecosystems, but few clues as to how these two systems communicate. Combing through journals and proceedings from conferences, it almost seemed as if agricultural researchers and human health researchers existed on two different planets. Sure, they often used identical language—terms like "microbiota" and "microbiome" (the collective DNA of the microbiota)—and many of the chemicals, nutrients, and bacterial species they described were exactly the same. They even published in some of the same journals and presented at the same conferences. But where were the studies that would help me figure out how human ecosystems mesh with those found on a farm?

Finally, after a slew of e-mail queries and Internet searches, I came across Justin Sonnenburg. Justin is a farmer of sorts, but a very different kind from Erick and Wendy. When he goes to work, he wears a crisp blue oxford shirt and pressed khakis, and there is no dirt under his neatly clipped nails. His "farm," within one of Stanford University's newer research facilities, consists of two hypersterile rooms filled with space-age inflatable plastic cubes. In these cubes, Justin and his collaborator and wife, Erica Sonnenburg, raise and study a strain of gnotobiotic mouse, a rodent that has no bacteria in its gut. For the Sonnenburgs, this "microbial nudity" offers the perfect opportunity to test what happens when microbes from one ecosystem are transplanted into another ecosystem.

The Sonnenburgs are part of a posse of young scientists who were groomed at Washington University in St. Louis by the guru of gut ecology, Jeffrey Gordon. When they started in Gordon's lab in 2003, little was known about the connection between the billions of bacteria in our intestine and our health, but over the next five years the field exploded. Largely owing to the work of Gordon and his team, we now know that gut microbes play a vital role in regulating everything from our metabolism to our immune response.

"It was like we were discovering a whole new organ," Justin told me during our first phone conversation. From his perspective, unraveling the mysteries of intestinal microbiota offers us as many therapeutic possibilities as sequencing the human genome.

"In fact, it may even have more potential," he added. "Gene therapy of our human genome is really tough to carry out, while microbiomes are more plastic, or more pliable. You can change a microbiome in a day—they're like a lever on our biology."

When I asked him for an example of this type of plasticity, he told me about a colleague's experiment that also involved the gnotobiotic mice. They transplanted intestinal microbes from an obese mouse into the gut of a lean mouse. Within days, the recipient mouse ballooned, even though there had been no change in its daily chow intake. Somehow, the new bacteria had served as the "lever" that altered the mouse's metabolism.

Even over the phone, it was easy to tell that Justin Sonnenburg was really excited about this bizarre experiment and that he belonged to that special camp of scientists who greet unusual, somewhat tangential questions with an open mind and genuine interest. He seemed like the perfect person to talk to about Allie, and so I decided to pay him a visit in Palo Alto.

The first thing I noticed as I entered Justin's sparsely furnished office was the image on his screen saver. Initially I thought it was a close-up photo of moss-covered pebbles in a dry creek bed, covered with moss, but on closer inspection I realized that it was a view of a human intestine through an electron microscope: what I thought were pebbles were actually colon cells, and the moss turned out to be sticky mucus glycans, a protective mixture of proteins and carbohydrates produced by the intestinal wall in concert with resident bacteria.

I told Justin about Allie, describing her digestive problems, her vitamin deficiencies, and other symptoms and asked him what impact the ecology of a farm like Jubilee might have on her health.

"This gets at the heart of some of the most important questions we are now dealing with in the field!" Justin replied, jumping in. He explained that we used to think of humans as closed systems, with the bacteria in our intestine being uniquely adapted to us and having little to do with the bacteria in our outside environment. Of course, microbes pass through transiently, as happens when we suffer from food poisoning or other infections, but by and large, we long assumed the two universes to be separate and our surrounding microbes to have little enduring influence on our health.

"But now that we can sequence an entire microbiome, using a technique called metagenomics, we're finally in a position to begin connecting the dots." He told me about his colleagues in France who had recently discovered some of the exact same DNA in both ocean-dwelling bacteria and the microbiome of Japanese people. Presumably the marine bacteria, which thrive on seaweed, hitchhiked their way into the human gut via sushi and other seaweed dishes and passed their seaweed-digesting DNA on to microbes living within the human host. The end result was that many Japanese—and possibly other people from seaweed-eating cultures—acquired a greater ability than the rest of us to extract valuable nutrients from their nori.

"This discovery is probably just the tip of the iceberg," explained Justin. "My guess is that we'll soon find out that microbes in soil and oceans are playing a huge role in our health. They do things for us that our own DNA is not capable of doing." In short, he seemed to think that this game of pass-the-gene is playing out all the time in each one of us.

Then, as he clicked through a series of PDFs, Justin told me about a groundbreaking study that illustrates just how intimately our microbiome (and our health) is linked to the natural world.

A TALE OF TWO DIETS

Before us on the computer was a cluster of round, thatched-roof mud huts. This study, Justin explained, was done by Italian researchers who took it upon themselves to compare the feces of two groups of young, healthy children, one group living in the remote village of Boulpon in Burkina Faso while the other lived in the researchers' hometown of Florence. Although both groups of children were examined and deemed to be "healthy," this term was a marker of true resilience for the Boulpon children, since they lived in a non-industrialized area where diarrheal diseases and other infections were common. Needless to say, the diets of these two clusters of children could not have been more different. Boulpon is a subsistence farming community where all food is locally produced, harvested, and prepared by the women in the village using a full-circle system with no outside inputs. The authors of the study surmised that these foods—minimally processed peas, beans, millet, vegetables, spices, the occasional free-range chicken and termites in the rainy season—were nearly identical to those eaten ten thousand years ago by the first African farmers. By contrast, the Florentine children ate a diet rich in meat and refined carbohydrates that was not so different from the standard American diet. (This is evidenced by the fact that rates of childhood obesity in Italy almost match those in the United States.)

When the researchers compared the bacterial DNA in the feces of these two groups of children, they discovered that their internal environments were as different as their eating habits and their habitat: The Boulpon kids had more bacteria from the phyla of Bacteroidetes and Actinobacteria than the Italians, and they also had a much greater diversity of bacterial species overall, including some types that are rarely found in the intestines of Westerners. Bacteroidetes and Actinobacteria have long, specialized sequences of DNA and do an excellent job of extracting nutrients from low-calorie, polysaccharide-rich foods such as unprocessed fruits, vegetables, and grains. Scientists are just beginning to understand the characteristics of these phyla, but Justin mentioned that they seem to play a role in de-activating unfavorable bacteria, cancer cells, and allergens. They also produce substances that repair intestinal walls and remove toxins. This protective function might explain why the Boulpon children were healthy and free of diarrhea despite the fact that there were plenty of noxious microbugs in their environment.

The children from Florence, by contrast, had many more members of the Firmicute and Proteobacteria phyla in their stool samples. These are the Homer Simpson bacteria. They thrive on the by-products of fats and sugars and predominate in the intestine when refined flour, dairy, and meat are the mainstay of the diet. They have a great capacity for what Justin calls "energy harvest": They package the calories from these foods and transfer them to the waistlines of their vertebrate hosts, whether cows, chickens, or humans. As you can imagine, this offers us a health advantage when calories are scarce but becomes a disadvantage when high-energy food is available in abundance. Firmicute and Proteobacteria also tend to be hardier, withstanding exposure to antibiotics, pesticides, food preservatives, and other chemicals that quickly damage less resilient phyla. Finally, these bacteria are more inflammatory, inducing an exaggerated immune response within their host. This makes perfect sense when you realize that Firmicutes and Proteobacteria phyla include the genus of Salmonella, Shigella, Klebsiella, Listeria, Clostridia, and E. coli, all capable of making headlines as the culprits in food poisoning outbreaks.

As Justin told me about these phyla, he reminded me about the gnotobiotic mouse study he had described over the phone, before my first visit. Not surprisingly, the transplanted bacterial samples that caused the thin mice to gain weight were proportionately high in Firmicutes and Proteobacteria. Newer research also links these bugs to other chronic health problems such as diabetes, inflammatory bowel disease, and cancer. (In the next chapter, we discuss why these bacteria don't always cause symptoms and how, in moderation, they might play a positive role within farms and humans.)

"Basically," said Sonnenburg, "even though most of us are eating in a modern way like the kids in Florence, our genome is optimized to work with an ancient microbiota like the one promoted by the foods in Boulpon." Indeed, the locally grown, largely plant-based diet eaten by the Boulpon is similar to that eaten by most of our ancestors and, with few exceptions, it is similar to the diet eaten by most surviving traditional cultures worldwide.

FIGHTING EXTINCTION FROM WITHIN

It all made sense. The Boulpon farm children, who were continuously dosing themselves with a variety of plants and plant-loving bacteria, might offer a snapshot of what all our microbiomes should look like in order to best serve the demands of our DNA and avoid chronic health problems. They also clued me in to what might happen to Allie if she entered an eco-farm cycle, whether in Burkina Faso, Italy, or the United States. The bacterial kingdom of her intestine would shift from one where hardy, pesticide-resistant, meat- and sugar-loving bugs ruled to one governed by a more complex and diverse set of organisms. The good news was that this might happen relatively quickly. Justin told me that some of his mouse research suggested that the equivalent of one McDonald's Happy Meal could produce unfavorable changes in an intestinal population, but that these changes were reversible if a Boulpon-style diet was reintroduced. (Of course, the mash he feeds his mice is not a perfect representation of what one might eat in Boulpon, and mice are not humans.)

I thought of Jubilee's soil, so alive that it could digest a cow pat in a matter of days, and wondered what other benefits the farm's bacteria (or DNA from that bacteria) might confer in Allie by hitchhiking through her intestine on a broccoli floret or a sprig of parsley. Would they de-activate unfavorable bacteria or allergens? Repair intestinal walls? Remove toxins? Or maybe reeducate her metabolism?

TEST-AND-REPLACE REVISITED

Justin had been very helpful, indulging my hypothetical questions about what would happen to Allie's "tilth" should she become a part of a biodynamic farming community. But Allie didn't live anywhere near a farm, and it was not likely that I was going to get her to move to one. So I asked Justin if he had other suggestions for how I might help her shift her microbiota and improve her health.

"My optimal solution," he continued, "would be to have a simple, reliable diagnostic test to identify a person's microbial status. Then we could give them an appropriate prebiotic or probiotic." Prebiotics are substances that are intended to nourish gut bacteria, while probiotics contain a dose of live microbes.

"But as you've learned, there are so many variables that must be considered, and so many unknowns. . . ." His voice trailed off.

Having spent this time learning from Justin, I immediately saw his point. Who was to say what was the "right" probiotic formula? Supermarkets, drugstores, and vitamin shops are filled with items that promise to propagate healthy bacteria and restore balance in our intestines. But if you read the product labels, you realize that no two products contain the same combinations or quantity of bacteria. What are these recipes based on? Research? Intuition? Each time researchers identify new bacterial subpopulations, such as the Xylanibacter in the Boulpons, do they reconsider what should be included in such a pill? As for stool tests that evaluate intestinal populations, what is the gold-standard "healthy flora," the norm against which these population profiles should be measured? Should it be healthy kids in Florence, or in Boulpon, or somewhere else? Finally, are stool samples the best reflection of bacterial life within the colon, or is a more invasive test (intestinal sampling via colonoscopy, for example) needed to get a true read on the situation?

"We're just starting to figure this out," conceded Justin. "It might be that even rare players, microbes that constitute less than 1 percent of our entire intestinal tract, have a disproportionately large function."

I left Justin's lab and wandered back to the parking lot, passing the perfectly manicured lawns and shrubbery that surround his research facility. A gardener was spraying something on the grass, and for a moment I couldn't help but imagine the scant, and likely homogenous, microbiota that existed under that medicated landscape. Eventually biomedical research might sort through this complexity and find the perfect tests and pills that could help rebalance Allie's bacterial flora and her overall health. But so far the most promising approach was being practiced by Erick and his like-minded colleagues who were using an eco-cycle to bring health to animals, soil, and community. How could someone like Allie become a part of such a cycle?

ENTERING THE CYCLE

I started thinking about John Romanelli, the woodworker with the heart of a farmer. He spent most of his days in a woodshop, and yet he felt connected, body and soul, to Jubilee. Not only did he do a half day a week of work-share, but almost everything he ate came from that soil and much of his socializing occurred on the farm. It was the same for the other "work-shares" I talked to during my internship. Then I remembered a conversation I'd had months earlier with chef and restaurateur Alice Waters. She told me about feeling that her restaurant, Chez Panisse, is connected to the farm cycle because each week Green String Farms delivers boxes of fresh produce to her restaurant in Berkeley and collects, in exchange, that week's kitchen scraps to add to the farm's compost.

Both Sir Albert and Rudolf Steiner wrote about people like John Romanelli and Alice Waters, who lived within the greater "farm community" and proposed that they derived health benefits from their association with the land. Interestingly enough, a group of European researchers (who appear in the next chapter) have recently collected data that supports this notion: They have shown that suburban children who attend Steiner schools have fewer colds, and suffer less from asthma or allergies. These schools are often located on or near biodynamic farms and most of them offer lunches and snacks sourced from the farms.

The next time I saw Allie I shared what I had learned at Jubilee and at Stanford. Given that nothing else was helping, I asked her if she would be open to joining a farm cycle. At first she stared at me blankly: She seemed to find the whole idea a bit far-fetched. But then she surprised me by saying that she was willing to give it a try, the only caveat being that it all needed to happen within the confines of the city. She just wasn't the type to swap her iPhone for a hula hoe and move to the countryside. Here is the plan we made together:

FIVE STEPS TO PERFECT TILTH

Step 1: Invest in Farm-Fresh

Yes, it was an investment, but Allie quickly realized that she'd been spending just as much money on frozen dinners, takeout meals, energy bars, and supplements. She thought that a weekly trip to the farmers' market would require too much planning and organization (and she found the whole "foodie" scene a bit precious), so she started to shop at a local market that bought directly from farmers without a middleman. She also signed up with a CSA (community-supported agriculture) program that delivered a weekly box of farm-fresh vegetables to a drop-off spot in her neighborhood.

From my time at Jubilee, I understood that the only way to become a part of a farm cycle is to eat food that has been grown in a sustainable, eco-cycle model; I also understood that the organic label is not necessarily a guarantee of that. In fact, when researchers have looked at the impact of organic systems on both soil quality and food nutrient levels, they have found that they often fare no better than conventional farms. Sustainable or biodynamic farming, on the other hand, seems to consistently score better on both measures. To better understand where her food comes from, I encouraged Allie to meet her farmers or read about their farms online and learn about their farming methods. (For a great resource to help you discover sustainable farms, CSAs, and markets in your area, visit www.localharvest.org.)

Here are the kinds of questions she could ask: What role do animals play on your farm? Do you import minerals and fertilizer or do you recycle your own fertility? Wendell Berry believes that the best question to ask is whether the farmer lives on the farmland. As you can imagine, when a farmer raises a family on the land and is literally rooted in the soil, he or she is going to be much more mindful about caring for that property. (Industrial farms are mostly owned by businesspeople who rarely set foot on their land acreage.) I also suggested that Allie take the time to visit her CSA farm since it offered quarterly farm tours for members. Who knows, she might make a friend or two in the process.

Step 2: Eat for Biodiversity

Impressed by the story of Boulpon, Allie set out to save her vegetable-loving microbiota from extinction. She realized that a diverse diet meant a diverse microbiota, so she exchanged her energy bars for a variety of primitive grains, including millet, sorghum, barley, faro, spelt, maize, and bulgur. She also began to let the seasonal herbs, fruits, and vegetables from the farm dictate what she put on the table.

Here is a list of some of her favorite farm foods: onions, leeks, garlic, basil, parsley, thyme, Jerusalem artichokes (sunchokes), honey, goat milk, kale, dandelion greens, spinach, broccoli, Brussels sprouts, purslane, squash, asparagus, carrots, tomatoes, blueberries, kiwis, cantaloupe, cherries, plums, apricots, apples, and oranges.

All of a sudden her diet got less monotonous, as it changed throughout the year.

It turns out that some of the foods that Allie had been told to avoid because they were "gas-producing"—including Brussels sprouts, broccoli, asparagus, leeks, and peas—happened to be the best prebiotics, meaning that they are the preferred diet for those beneficial Bacteroidetes and Actinobacteria phyla. These same foods also offer a rich supply of antioxidants and vitamins in a form that is safer and more easily processed than supplement formulations. As Allie began to eat these foods, she did notice that they made her a little gassy. But if she avoided eating too many of these foods at once, they were easy enough to digest.

Allie also swapped out conventionally raised chicken and other animal products for those that had been sustainably raised. Research shows that meat from these animals has a better nutritional makeup and fewer antibiotics and hormones.

Step 3: Eat Dirt and Bugs

Well, not literally . . . but I encouraged Allie to not be too compulsive about scrubbing her farm-fresh produce. I reassured her that getting a little bit of soil in her system from food grown in healthy soil would be just fine. Who knew what beneficial bacteria and minerals might be coming along for the ride? I don't mean to suggest that food-borne illness is not a real concern, even in the United States and other Westernized countries. In fact, one in six Americans get sick from contaminated food every year, generating billions of dollars in health care costs. On average, however, fewer than three thousand people per year die from these infections, and if you comb through the reports of the Centers for Disease Control (CDC) on these *deadly* outbreaks of E. coli, shigella and salmonella, listeria and botulism—as I have—you realize that well over 95 percent of the food involved is grown, processed, and packaged by large commercial manufacturers. Of course, the denominator is larger for processed food, since national brands still make up the lion's share of the American diet. But even on a per-serving basis, the risk of infection from eating industrially produced food is so much greater that during the 2006 E. coli–tainted spinach scare, Dr. David Acheson, who was in charge of food safety at the Food and Drug Administration (FDA), counseled Americans to avoid the illness by sticking to local produce. He told the *New York Times:* "Clearly the risk is significantly reduced if you know the farmer and know his farm."

Along these same lines, I encouraged Allie to eat the outer leaves and the peels of her carrots, cabbage, Brussels sprouts, artichokes, apples, and other produce. In general, this tougher, pest-nibbled, sun-exposed covering is precisely the part of the food that has the highest concentration of nutrients and is best suited for feeding those good Bacteroidetes and Actinobacteria in the intestine.

I also mentioned to Allie that while her stomach pains and other digestive issues might improve by taking probiotics or bacteria in pill form, emerging data show that some bacterial supplements offer little benefit and can sometimes pass on antibiotic-resistant genes to other intestine-resident microbes. Fermented foods, on the other hand, can offer an excellent alternative as this source of bacteria is often more biodiverse and more aligned with our health needs. As Justin Sonnenburg explained to me, our ancestors, who for millennia had no access to preservatives, pasteurization, or refrigeration, evolved to tolerate the bugs found in their rotten food and possibly to reap health benefits from them. In our modern lives, our daily brush with putrefying bacteria is minimal to nonexistent; foods produced by controlled fermentation come the closest to offering us a collection of bacteria and yeast that matches that found in our predecessors' diets.

I'm especially impressed with the non-dairy fermented foods, since these grow more bacteria that belong in the Bacteroidetes and Actinobacteria phyla. These foods include fermented pickles and cabbage (sauerkraut and kimchi), barrel-aged vinegar, and fermented whole soy products such as miso and tempeh.

Step 4: Don't Kill Your Good Bacteria

Allie also came to realize that the preservatives and chemicals in her food, as well as her high-dose supplements and acid blockers, might be obliterating beneficial bacterial species. She had also taken many rounds of steroids and antibiotics in the past, two classes of medication that have been shown to foster less beneficial intestinal microbiota. (Several studies by David Relman at Stanford have shown that recurrent antibiotic use may result in a permanently altered gut ecosystem for some individuals.) Of course, there are times when a drug is necessary, but Allie now realized that some of her doctors had been way too cavalier about prescribing these powerful medications, especially for long-term use.

Step 5: Engage in Farm Love

The close community at Jubilee made me a little envious. Sure, there was bickering, but I've rarely spent time in a place where I felt more warmth and interconnection. I brainstormed with Allie about how she could find this kind of community in San Francisco. She looked into volunteering in a local park and a community garden, but eventually decided to spend a couple hours each week in an elementary school's garden. During a recent office visit, she told me how much she enjoyed the students, teachers, and volunteers. Months later, after spending time farming in the Bronx, I would come to understand how farming, in addition to building community, can act as an antidepressant. Here are some explanations for this mood-elevating effect:

- Gardening increases the chance for interpersonal connection and boosts one's sense of purpose.
- Gardening is a good form of exercise, as it includes weight-bearing, deep squats, core strengthening, and lots of walking, and physical activity is a powerful antidepressant.
- Like plants, we fare better when we spend time in sunlight. Gardening outdoors helps raise our levels of vitamin D, a pro-hormone needed to make the antidepressant neurotransmitter, serotonin. I know that some dermatologists and the skin product industry have turned "sunshine" into a dirty word, but remember that most skin cancers are caused by sunburns or prolonged exposure to sunshine (and some might not be related to the sun at all). Now that Allie was spending time outside, she always wore a hat. But she made a point (weather permitting) of getting about fifteen minutes of sun on her arms and legs. After that, she would cover up . . . like a farmer.

LEAVING JUBILEE

My last day at Jubilee, Erick treated all the farmhands and me to lunch at the local pizza parlor. Among other things, we shared a pitcher of soda and an Aloha pizza with canned pineapple chunks, packaged ham, and lots of glistening white cheese—all served on a fluffy crust. I didn't think much about it, but apparently my stomach did. As we drove back to the farm, I suddenly felt like someone had given me a swift punch right above my belly button. It dawned on me that I'd been mainly eating from the farm for the previous two weeks, and that this last meal represented a dramatic excursion out of that cycle. (No doubt all the fat- and sugar-processing Firmicutes in my gut were high-fiving and having a holiday.)

That afternoon I thanked Erick again for lunch and jokingly told him that the prospect of leaving the farm had given me a stomachache. He laughed. Without thinking, he reached down and grabbed some of his soil and sifted it through his fingers. "Yeah, this place is pretty great. I don't like to leave it either." It was then that I noticed that he was wearing one of his favorite T-shirts, emblazoned with the slogan: COMPOST HAPPENS.

POSTSCRIPT

About four months after my stay at Jubilee, I called Erick to say hi and ask him some more questions. I told him that I was going to write about his test-and-replace approach and his conversion to biodynamic farming.

"You know," he said, "I'm actually thinking of testing again this year. Not because I think there's anything wrong with my soil. I think it's great. But it's more that I want to prove to myself that my biodynamic system is working." I understood exactly what he meant. A couple of weeks later, I once again saw Allie in my office. She had stopped almost all her pills except for a low dose of her antidepressant and was feeling well for the first time in several years. She said that she loved being a part of a farm eco-cycle. I was delighted for her, but of course I could not help but wonder which part of this whole experiment had truly made the difference. Was it the microbes? The nutrients in the food itself? The new friends she had made through her farm work? The new sense of purpose? The physical labor? The sunshine?

The fact that she was no longer grieving so intensely for her father? Or that she had abandoned all those pills? Or maybe all of the above, and some things I hadn't thought of? After all, each of these factors played a role in the complicated web of Allie's life. I remembered my recent conversation with Erick and proposed to Allie that she too might get some follow-up labs. Not the extensive testing that had been done earlier—I just wanted to measure some standard nutrients such as vitamins B_{12} and D and iron, all of which had been low previously. At first she hesitated. She felt great and didn't want to go looking for trouble. But soon after, curiosity must have gotten the better of her, because she called and asked for a lab slip. A few days later, we added one last page to her bulging manila folder.

It didn't surprise me that this time everything was normal.

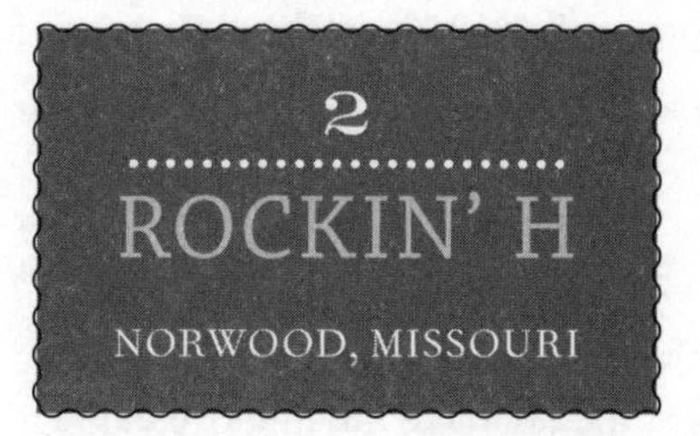

Raising Kids Bison-Style for Maximum Resilience

It is important to consider "the forest" of changes that occur with westernization, as well as the specific "trees"; and that the package of changes that come with westernization and increased hygiene may increase asthma risk.

—Brooks, Pearce, and Douwes, "The Hygiene Hypothesis in Allergy and Asthma: An Update" (2012)

ON ERICK Haakenson's bookshelf, I came across a pile of past issues of *Acres USA*, a monthly magazine whose tagline reads, "The Voice of Eco-Agriculture." Given the publication's progressive bent, I was not surprised to discover that a certain proportion of the articles tended toward the New Age and the esoteric, with titles like "Paramagnetism and Plant Vitality." By contrast, each one of Cody Holmes's regular contributions from

Rockin' H Ranch was peppered with wry cowboy humor and dished out the kind of useful, down-to-earth advice that could be garnered only from decades of hands-on experience as a Missouri cow-calf farmer.

"Remember, if your farm resembles a golf course, you would probably make more money off your farm through green fees than selling milk or meat," he cautions in one article about dairy grazing in which he extols the nutritional value of weeds and advises his fellow farmers to avoid herbicides. And in an essay that explores the merits of leasing versus buying farmland, he opines: "Allowing a bunch of old cows to pay off the debt on your own farm will someday have much more value to you than contributing to your neighbor's farm debt through a monthly lease."

I enjoyed Cody's practical musings so much that when I returned home from Jubilee, I decided to explore his website. There I discovered that this no-nonsense cowboy likes to refer to himself as a holistic rancher. Now I was really curious. What could have prompted a man who regularly blogs about "those liberals from California" to come up with such a touchy-feely moniker? At this point I was intrigued enough to give him a call and ask why he chose his "holistic" handle.

"You're from California, and I don't mean you any disrespect, but I know they have a lot of do-gooder fruits and nuts out there," he said in a soft Missouri drawl. "What you need to understand is that I've been a typical rancher, rodeo rider, my whole life—the opposite of a tree-hugger. For me, economics has always been number one."

"What's funny is that my wife Dawnnell and I have morphed into rancher, foodie, tree-hugger, worm-lovers," he chuckled. "Economics is still number one, but we've realized that to be more economical, we have to be holistically minded conservationists."

Cody began to describe how the change had taken place. He had grown up on a dairy farm, and at seventeen he had started his own small beef operation with a handful of cows that had been his 4-H project. Over the next three decades, his farm, Rockin' H Ranch, grew until it included more than a thousand acres and twelve hundred Angus beef cows. But as hard as he worked, Cody's expenses (or what he referred to as "inputs") far outpaced his profits and he was unable to support his family just by farming. Like most of his neighbors, he needed to work off the farm—in his case, doing accounting for other businesses—to make ends meet.

"I was working like a dog and spending more and more money on hormones, vaccines, dewormers, insemination fees, corn feed, nitrogen fertilizers, and tractor fuel, and yet the commodity price for beef had not changed in twenty years."

At the same time, he noticed that his land—his most valuable asset—was slowly degrading. Each year more and more topsoil washed into the nearby river, and the water table dropped to an ever lower level.

"But even before all this became prohibitive," he continued, "it was my conscience that poked me in the head." I smiled, imagining Cody's conscience as a giant cow prod, jabbing him in the ear.

"We used to run the cattle through the squeeze chute[1] and use Ralgro hormone implants. These are pro-growth hormones. We would have one for the boys and one for the girls. For years we were running the cows and doing the injections of antibiotics and growth hormones, and I was keeping one out to be injection-free for my kids."

Then, about ten years ago, Cody met his second wife and co-farmer, Dawnnell. Like Wendy at Jubilee, she served as the catalyst for radical change on the farm. She and Cody decided that enough was enough. Why was it okay to feed these cows to other people's kids and not their own? So they stopped buying most of the usual farm inputs.

"We quit cold turkey," said Cody. "Quit purchasing soil amendments and animal feed and quit injecting everything into our cattle."

Around this same time, he became inspired by the writings of Andre Voisin, an early-twentieth-century biochemist and farmer who showed that soil health is directly linked to animal and human health, and by Allan Savory, a farmer, biologist, and environmentalist who is a noted pioneer in the practice of holistic land management. "I realized that for things to really work I needed to redesign my entire farming system. That way I could employ the only free labor I had at my disposal, those billions of bacteria and other microbes in the soil. My goal was to have their collective weight underground be greater than the livestock above, and for them to give me healthy soil and plenty of grass to feed my animals twelve months a year."

I told Cody that, because of my time at Jubilee, I was well acquainted with those microbes and their benefits and asked him how things were going with his own farm experiment.

"Pretty good," he answered. Now, ten years in, he felt that it was really starting to pay off. He had healthier soil, the water in the surrounding creeks seemed clearer, and he and Dawnnell were making a good annual income as farmers. He characterized his whole operation as having lower production but higher profits. He had pared his herd down to seven hundred, but was able to earn more per pound of meat since he relied on free worm-labor and no longer purchased inputs. All this gave him more time to advise struggling farmers, develop a community on Facebook, and talk to curious people like me. But what really got my attention was what Cody said about his cattle.

"My cows are healthier than they've ever been. These days I have no problems with infertility, or 'open cows,' as they call it in the industry, and no major illnesses with the calves. I see none of the infections and respiratory diseases that routinely plague corn-fed, barn-raised animals. My vet costs are minimal, and I have no calf deaths."

"Really?" I said, sounding dubious. I could still recall how frequently, as a child in upstate New York, I would stumble across a decomposing calf in our neighbor's field, swarmed with flies and turkey vultures.

"Zero!" said Cody emphatically. "The folks from the university say we're lying, so I use our ranch as a teaching tool. I invite you to come out and see for yourself."

The following spring I took Cody up on his invitation. I wanted to meet him and find out what was making his cows and calves so resilient.

ROCKIN' H RANCH

Heading east along Missouri's Route 60 toward Rockin' H Ranch, I was delighted to discover a landscape of rolling, lush pastures dotted with grazing black Angus cattle. Somehow this was not the Ozarks that I had imagined. Maybe it was that bleak movie *Winter's Bone* that had given me a false expectation; set in southern Missouri, it paints a picture of broken humans and crumbling meth shacks. Or maybe it was Cody's response to my question about why he grew cattle rather than vegetables.

"Have you ever been to the Ozarks?" he asked politely. "If you saw our thin soil, you'd understand why we grow good cows but a carrot doesn't have a chance."

Not long after passing a sign for the Laura Ingalls Wilder home and museum, I found Cody's exit and started northward, first on a narrow country road and then, for about a mile, on a rubbly dirt one. It was just turning dark when I finally pulled up to his modest, one-story farmhouse. I walked through the yard and was greeted on the front stoop by a friendly calf who seemed to think she was a dog. Dawnnell opened the front door.

"Oh, don't mind her," she said. "This one was rejected by her mom, and we're bottle-raising her."

Dawnnell had just finished the evening milking of her twelve Jersey cows, and her hands were wet and sudsy from washing out her equipment. She looked young and capable; something about her reminded me of Debra Winger in the 1980s movie *Urban Cowboy*. I followed her into a cozy living-dining room, where the wooden table was already covered with plates of food: grilled steaks, a big green salad, and macaroni and cheese. Taylor, her sixteen-year-old daughter from her first marriage, was at the table along with another teenager, a farm intern named Faith. Cody sat in the far corner wearing a T-shirt and overalls, his hair matted down, presumably from a day spent under the cowboy hat that rested on a nearby chair. He too looked much more youthful than I had expected, based on the fact that he mentioned grandchildren during our phone conversation.

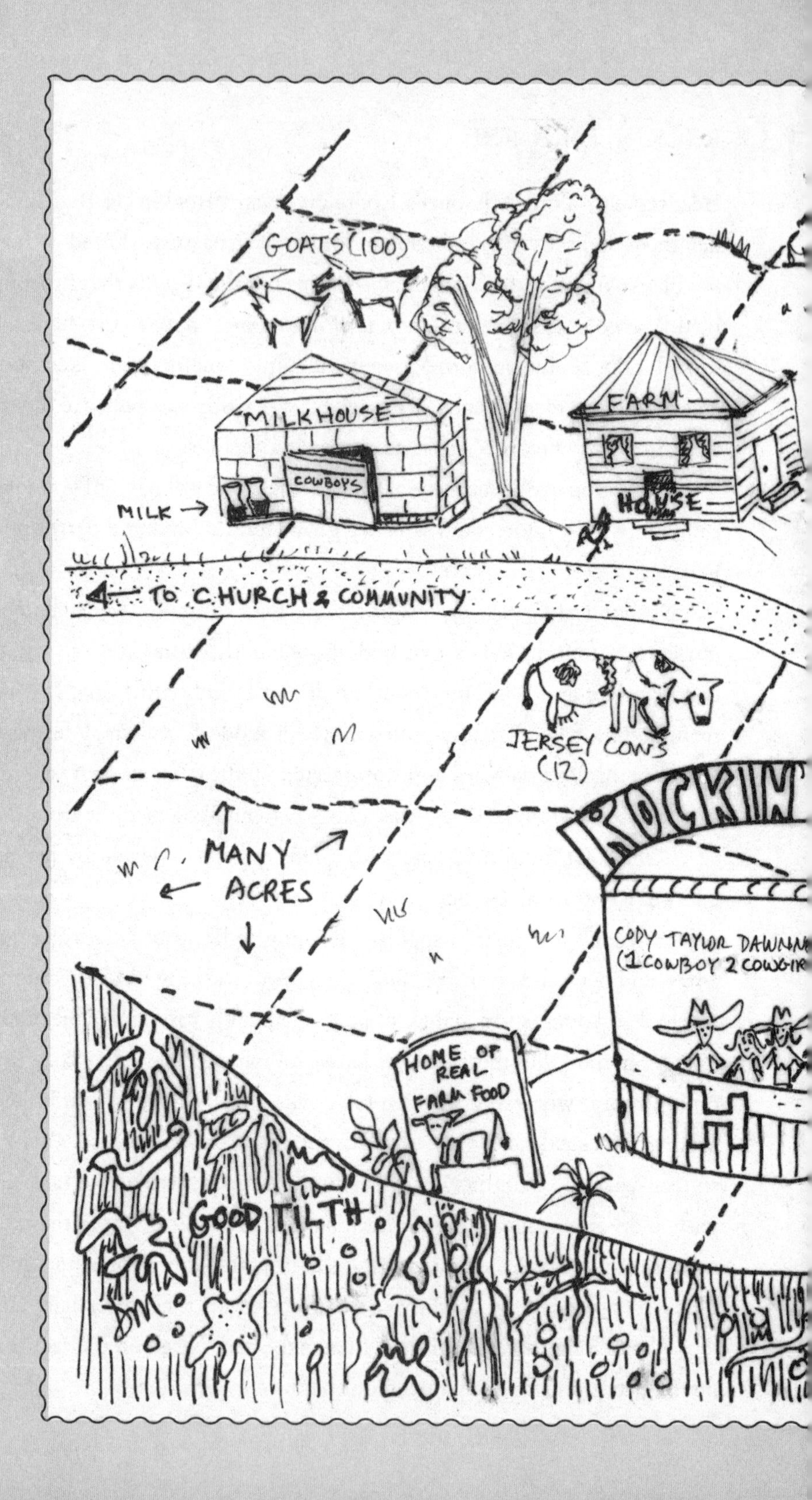

GOATS (100)
MILKHOUSE
COWBOYS
MILK →
FARM
HOUSE
TO CHURCH & COMMUNITY
JERSEY COWS
(12)
MANY
ACRES
CODY TAYLOR
HOME OF
REAL
FARM FOOD
GOOD TILTH

N↑

SHEEP (500)

HORSES (2)

TO RIVER→

SUMO COWS (700)

CALVES

(200)

FORBS

HENS (300)

CODY'S ATV

PIGS (150)

CURTNER RD

TO NORWOOD & COMMUNITY

COOKHOUSE

COWBOY GATHERING SPOT

GARDEN

OBSOLETE FARM EQUIPMENT

SQUEEZE CHUTE

Rockin' H Ranch

"You made it!" he said, not bothering to hide his surprise. I guess he half-expected that the nutty Californian, true to her kind, would flake out at the last minute.

We bowed our heads while Cody said grace, and then we focused our attention on the homegrown steaks that Dawnnell had placed in the middle of each plate. Like all the farms I visited, the food itself served as a kind of shorthand for everything that took place on that land . . . and these steaks were delicious. We exchanged small talk for a while, and then Cody pushed his plate aside and asked me what brought me all this way.

I repeated the guiding questions given to me by Wendell Berry. What was here before he came? What did nature require of him here? What did nature help him do here? And then I added, "Basically, I want to learn your secrets for raising healthy cattle."

Cody nodded and started to tell me about a diary he'd once read that was written by an early-nineteenth-century pioneer. At first, I had no idea what this story had to do with the questions I'd just posed, but Cody struck me as someone who rarely wasted his words on trivial matters. So I sat back and listened.

"The pioneer had just made it across the Mississippi with his wife, a baby, a horse and wagon, and a couple cows," he said, moving his steak knife along the table as if it were an advancing settler. "In the distance, they could hear thunder." He patted the tabletop with his free hand to add some sound effects.

"Then they came up on a rise, and guess what they saw?" At this point, Cody's voice turned a little theatrical. "Bison. There were bison as far as the eye could see.

"Those settlers cut right through them." Cody sliced the air with his knife, and I imagined that brave couple and their tiny baby making their way through a thunderous gathering of beasts. "They rode for three days," he continued, "and all they saw was bison."

"Then, on the fourth day, they moved beyond the herd and came across utter devastation. Those animals had stomped the earth and cleared everything in their path. But this was okay. You see, those cattle would move through four states grazing before they made it back to this spot, and by that time everything would grow back. In the early 1800s, before they were all exterminated, we had more bison than we now have cattle. And they did not need grain."

Cody explained that the migratory pattern followed by those bison has inspired his farming model. Sometimes he refers to it as "holistic grazing," or "mob grazing," but mostly he calls it "bison style." It is, in essence, a larger-scale version of the rotating pastures that Erick uses for his tiny herd at Jubilee. Instead of grain feeding or the low-density grazing that I saw earlier along the highway, Cody has returned to a natural system that predates beef farming in the Ozarks. Now his cattle move, en masse, through a series of small paddocks where they forage intensively and then deposit their meals, well processed, back into the soil. Since at least four or more months elapse before they move back to the same paddock, the flora and fauna have plenty of time to regenerate. This system allows Cody to run three times the number of cattle per acre as the average ranch in Missouri, and it is this system, as I would soon learn, that has allowed him and Dawnnell to forgo expensive farm inputs. But most important, bison-style grazing has made Cody's herd some of the healthiest cows in Missouri.

"If you want the easy answer to why they are healthier," he said, setting down his knife and flashing me a charming grin, "there isn't one. Each and every cog has some effect on the other parts that make up the whole. I can't just teach you how to build a fence or move cattle around. You have to understand that you need to change the whole environment to keep those cows healthy. That is what I call 'holistic.' "

I understood that I needed to be patient, pay attention, and pick up clues as I went along.

CLUE #1: IN A GLASS

It turned out I didn't need to wait long.

After clearing the plates, Dawnnell returned with dessert: a tall glass of milk. Even in the soft light of the room I could see that this was not your standard dairy product. The liquid was a beautiful buttercup yellow, and it clung to the glass in a way that store-bought milk—even whole milk—just doesn't do. Watching Cody, Faith, and Taylor drain their servings with a satisfied smack, I so wanted to do the same. But a couple of things were holding me back.

First of all, I am past the denial phase when it comes to my lactose intolerance. Dairy, especially when it is unfermented and drunk late in the day, expands my gut with such speed and intensity that I am left with little choice but to remove myself from polite company and find someplace private (and preferably soundproof) to spend the rest of the evening. Second, given that it was fresh from the cow, I could only surmise that this stuff was raw. Now, I have a number of patients who are completely gung-ho about raw milk, and I too drank it often on our family farm in upstate New York. I have fond memories of standing in the neighbor's milking shed, the air humid from the hot milk and nearby biomass of cow, and dipping my one-gallon stainless steel bucket into the churning tank. I loved that milk and never for a moment had the impression that it was unsafe to drink. But sometime in the intervening decades I'd adopted the same opinion as most of my medical colleagues: that raw dairy is a potential vector for all sorts of diseases and consuming it is foolhardy. Truth be told, I have never taken care of a patient whose illness could be traced to raw milk, and the Centers for Disease Control's food-borne illness statistics suggest that, on a per-serving basis, one has a ten times greater chance of getting sick from commercial deli meats than from raw milk. Still, there's that specter of campylobacter, salmonella, shigella, listeriosis, or, God forbid, E. coli O157, that I cannot shake. At this point, I'd come to regard drinking unpasteurized dairy as tantamount to having unprotected sex; in both situations, one should only consider doing it after a good deal of intimacy and trust have been established and preferably some testing has taken place.

Dawnnell clearly sensed my hesitation.

"This is fresh from the Jersey cows," she said by way of reassurance. Cody held his empty glass up to the light, stating that raw mother's milk was the only food his calves got until they learned to forage on grass, and he was certain that this was one of the reasons they were so healthy and disease-free. Then he and Dawnnell redirected the conversation from calf health to human health. Alternating sentences, they recounted story after story about sick people from the surrounding community who regularly drove up to Rockin' H to buy a quart of fresh milk. There was the colicky baby who had eczema and what Dawnnell referred to as "a whole bunch of gut issues." The child's sage grandma said, "This kid is in pain," so the mother started cutting things from his diet. When she eliminated dairy, his symptoms disappeared. But, explained Dawnnell, as he got older he asked for milk. They tried raw milk from cows that had been raised on grain, and once again, he fell ill. Finally, they gave him raw milk from Dawnnell's grass-fed Jersey cows, and he did fine.

Then there was the guitarist in their church choir. That guy had terrible irritable bowel symptoms for years, but his pain and bloating disappeared when he switched over to Rockin' H dairy. And how about the wife of a local natural healer? She had "all kinds of things wrong with her" until she became a regular customer.

As I listened to their stories, I thought about all the patients in my practice with similar health problems. Nothing brought kids in more often than allergic diseases like eczema and asthma, and not half a day goes by when I don't see a patient, like Allie, who has "all kinds of things wrong," including irritable bowel symptoms.

"At first we thought it was hocus-pocus," added Cody. "We couldn't care less if we ever sell a gallon of milk. We have over seven hundred head of cattle, and Dawnnell has just twelve dairy cows. But then we started to see people get better, and we started to wonder if there really was something to all this."

THE SENSIBLE PROFESSOR VON MUTIUS

Not long after my visit to Rockin' H, I was standing with my suitcase in front of Maistrasse 11 in downtown Munich, staring up at the tall spires of what is now the Munich Women's Hospital. I was in Europe for other reasons, but decided to make a detour to Germany to visit Professor Erika von Mutius, a physician whose asthma and allergy research unit was housed in this building. As I rolled my suitcase past ornate columns and down the marble hallway toward her office, I reviewed my agenda: I needed to get the real story about the health value of raw milk and to find out why, based on her research, farm kids are so healthy.

I became aware of von Mutius's research roughly two months before my visit to Rockin' H, when one of her articles, titled "Exposure to Environmental Microorganisms and Childhood Asthma," appeared in the *New England Journal of Medicine*. I stumbled upon the report quite by accident, but much to my delight, I realized that this study took place on farms—more specifically, traditional farms in central Europe. A farm would be an unusual setting for any pediatric health study, but especially for one written up in such a respected publication! I e-mailed Erika von Mutius, introducing myself as a fellow physician who was interested in the connection between health and farms. Could I pay her a visit?

So that is how I found myself on Maistrasse 11, shaking hands with the professor. It was immediately clear that everything about Erika von Mutius was sensible—from her fuss-free short hair and sturdy flat shoes to her reasons, as an academic investigator, for spending so much time focused on farms.

"The reason I started to do this is very simple," she explained in impeccable English, motioning me toward a chair in her roomy, high-ceilinged office. "In the late 1990s, a colleague of mine who also studies allergies and asthma was talking to an internist who works in the Swiss village of Grabs. The internist happened to mention that he hardly ever sees farm kids who suffer from the kinds of health problems that she studies."

The allergist colleague passed this observation on to von Mutius, and it piqued her curiosity. She started to look at asthma and allergy rates in nearby farming areas in Bavaria and Austria and discovered that they were impressively low. Meanwhile, she was seeing plenty of allergy and asthma in her pediatric practice in Munich. Naturally, she wondered if there was anything to be learned on farms that could help her urban patients, and this question eventually led to the creation of an extensive, multinational research collaborative.[2]

I asked von Mutius to explain what her group had discovered about the factors that protect farm kids.

"It's like this," she said, taking out a clean sheet of paper and drawing a flow chart.

"On the top level, the most important factor is the farm," she explained, writing "farm" at the top of the page and underlining it several times to emphasize this idea. "You should know that we don't have big industrialized farms in the areas we have studied, mainly small traditional farms."

Then she drew a pair of lines from the word "farm," dividing this overarching idea into two subcategories. One, she labeled "milk."

"We have studied the many exposures that these children experience, and it seems like drinking farm milk is an important one," she said, authoritatively tapping the paper with the tip of her pen. I asked her what she meant by the term "farm milk," and she gave me a sidelong glance.

"Of course this means raw," she said, adding the word "unprocessed" to her flow chart. In other words, milk that was practically identical to what Dawnnell had served me on Rockin' H: fresh from the cow, unpasteurized and unhomogenized.

FARM MILK RESEARCH

If you delve into the world of raw milk on the Internet, you find references to von Mutius's research on many different sites, from home-birth chat rooms to a Paleo Diet blog. When I mentioned this, Erika looked somewhat surprised and then laughed, clearly appreciating the irony that she would have such a crunchy fan club.

"What those people need to know," she said, "is that I am in no way convinced that raw milk is the answer."

But hadn't she just written on her flow chart that raw milk was one of the reasons those farm kids were so healthy and allergy-free?

"Those children in my study are getting the milk fresh from the cow," she said. "If they don't drink it in the morning, they throw it out and take fresh milk in the afternoon."

By contrast, she explained, most people who live off the farm are drinking raw milk that is a day old or older. This lag time, as well as further handling in the bottling process, allows all kinds of dangerous microorganisms to flourish. The bottom line, as von Mutius saw it, is that she's taken the Hippocratic oath to do no harm: If she promotes raw milk and it causes just one deadly case of E. coli–associated hemolytic uremic syndrome, then she will have broken that oath.

But what is interesting about Erika von Mutius is that her reservations about raw milk have not prevented her from studying how it might give farm children a health advantage. I view this as true scientific bravery since, in most biomedical circles, discussing any benefits to raw milk is practically taboo. This is especially the case in the United States, where the well-publicized position of both the CDC and the FDA, the two leading government organizations that evaluate food safety, is that drinking raw milk is extremely dangerous for your health. Given their unequivocal stance, it is easy to understand why Erika von Mutius has few collaborators on this side of the Atlantic, even though plenty of raw milk is consumed here. (I seriously doubt that the *New England Journal of Medicine* would have accepted the professor's paper had it mentioned "farm milk" as one of the factors protecting the farm children.)

On my flight to Munich, I had read several articles on raw dairy coauthored by Erika von Mutius. "Despite the notion of early milk consumption being a risk factor [for allergy and asthma]," she writes in one published in the *Journal of Clinical and Experimental Allergy*, "there is an increasing body of evidence that unprocessed cow's milk does not increase, but rather may decrease the risk of asthma, hay fever, and allergic sensitization." Her papers go on to explore the ways in which farm milk offers protection and, conversely, how homogenized and pasteurized milk might cause an allergic response.

Raw milk contains whey proteins and cytokines such as lactoferrin and TGF-Beta, which von Mutius and her colleagues believe are effective at boosting defense cells and fighting infection.[3] At the same time these factors act as buffers within the immune system, keeping it from overreacting to outside allergenic proteins or other invaders. (In many instances, it is our immune system's overreaction, rather than the foreign substance itself, that produces all the intolerable symptoms that we associate with asthma and allergy—the itch, wheeze, sore throat, and stuffy nose.)

On the flip side, the process of pasteurization, which entails heating to 165 degrees Fahrenheit to neutralize proteins and kill bacteria, can degrade the milk's protective factors. It can also cause precipitation of proteins called beta-lactoglobulins. These proteins organize into larger, less digestible molecules that can trigger allergy symptoms—especially when they're consumed by an allergy-prone individual.

Homogenization has its own potential problems. Here, milk is passed through a series of narrow high-pressure tubes that break the big floaty fat globules into evenly dispersed smaller ones. This treatment makes milk universally palatable by turning it from yellow to white, eliminating any bovine flavor, and creating a smooth, uniform liquid with no cream floating on the surface. (The cream on top is what I personally love about unhomogenized milk.) The high-pressure churning during homogenization also has the unintended effect of displacing allergy-causing proteins: Instead of remaining safely tucked inside the fat globules, they now lie on the membrane surface where they're likely to set off an inflammatory response within the human immune system.

Having read all this, I could understand Erika von Mutius's raw milk dilemma: On the one hand, as she has argued, the way we produce and process this most popular of foods has the potential to cause health problems—especially in those children who are already at risk for allergic diseases. On the other hand, although fresh-from-the-cow raw milk might confer real protection, it is hard to come by and, in her opinion, too risky.

"But there are practical solutions," she said obliquely. "And now I am devoting most of my energies to this." She was working with Swiss and German dairy researchers, and I think I heard the word "filtration," but then she said something about patents and I realized she was not offering any more information on the subject.

GARBAGE IN, GARBAGE OUT

I don't know whether Cody and Dawnnell are part of the crowd that follows von Mutius's research, but after visiting Munich, I realized that they agreed with her on every point. Thinking back to that first night at their dining table, I realized that they had both mentioned the disruptive effects of heating and homogenization and the protective power of undisturbed whey. They were also aware that their Jerseys, who lived on grass twelve months a year, produced a milk with a healthier fat profile than the dairy products sold at their local Walmart.

"How can you expect to get something healthy from your average industrial dairy cow?" Cody asked me as I bravely took my first swig of Dawnnell's delicious offering. "It's garbage in, garbage out. Many cows are living almost up to their bellies in feces and have bad nutrition. Their pasteurized milk is not fit for human consumption. Yes, you should pasteurize their milk—and then feed it to the hogs. That is, if you don't like those hogs."

But then I asked Cody whether he would ever feed his calves or his family someone else's raw milk, and his mouth twisted in distaste.

"Well, no, why would I do that?" he asked. "Who knows where that milk has been?"

CLUE #2: A MICROBIAL SWAP MEET

My first morning on Rockin' H, I woke up energized. To my amazement and relief, the storm I had expected in my intestine never happened. In fact, there was not even the faintest rumble and certainly no sign that I had ingested a terrible E. coli–like pathogen. Feeling as if I'd dodged two bullets, I admitted to myself that Dawnnell's milk was one of the most delicious things I'd ever tasted and resolved to tempt fate and try another glass at breakfast.

The sun was just coming up, and although the spring day promised to turn warm, it was still bitterly cold outside. I wandered into the milking shed in search of heat and Dawnnell and found her bent under a tawny Jersey named Josie, attaching a tentacled electric milking apparatus to each of the cow's teats. Soon the milk began to rhythmically flow through the transparent tubing and into the stainless steel milking can.

Dawnnell stood up and lovingly patted Josie's flank. This girl, she cooed, was over eight years old. Echoing something that Cody had mentioned the night before, she told me that the average cow in a conventional operation milks for one and a half cycles, or until she is about three, then her milk production begins to drop and off to the butcher she goes. On the other hand, Rockin' H cows like Josie continue to give a steady two and a half gallons per day well into their tenth cycle. Dawnnell was certain that their holistic approach to farming had boosted her Jerseys' production and extended their life span.

The milking shed was an adorable whitewashed building and Dawnnell maintained it with a higher standard of cleanliness than most people impose in their kitchen. Everything was put away with not a cobweb in sight, the cement floor was spotless, and the air smelled like fresh grass. Quite different from the dairy barn where I spent time as a child that was spattered with manure and often had a sour odor. I shared my observation with Dawnnell, who said that it was the fresh grass diet that made the difference. Cows fed soy, corn, and other processed grains have looser bowel movements—and things get much smellier. Interestingly enough, one might make the same observation in humans.

As we chatted and milked, Josie's gangly calf stumbled in through the half-open door and gave her mother's rump a good sniff. Next, sticking out a velvet tongue, she inched forward and licked Josie's udder right above the milker attachment. Almost on cue, Brutus the dog appeared in the doorway and treated the calf to a similar rear-end sniff while two barn cats, their fur dappled with morning dew, slunk past, almost grazing the rack of empty quart-size Ball jars that would soon hold Josie's milk. Suddenly I saw this gleaming milk house in a whole new light; far from being an antiseptic chamber, it was a petri dish teeming with microbia. Everywhere I looked, worms, viruses, bacteria, and fungi were being deposited and exchanged.

Right then, I kept these insights to myself, but later that morning I could not help sharing them with Cody as we headed out to the pastures. I was riding shotgun on his four-wheeler—the only piece of "heavy-duty" farm equipment he's needed ever since he gave up all external farm inputs and let his animals migrate bison-style.

"That's exactly right," he shouted over the noise of the engine. "I believe in my bacteria. It's what keeps this soil healthy, so I get healthy animals and a healthy farmer."

Then, a few engine strokes later, he added, "My daughter is a nurse, and I love her to death, but she is totally antibacteria. She uses all those soaps and hand sanitizers on her kids, and they get sick all the time. Sometimes I just want to take those kids out and feed them manure."

Cody idled the engine and briefly hopped off his machine to open a barbed-wire fence. Then we charged into the seven-acre paddock, which looked as if it had recently hosted a big herd of cattle but now was empty save for a mobile henhouse. He explained that normally a vacated field would be thick with flies, attracted to the fresh cow pats, but that the larva-eating, manure-dispersing hens served as excellent bug control. As we approached, dozens of multicolored birds streamed out of the trailer doors and swarmed us like fans on a rock star's limo. One even made her way onto the crown of Cody's cowboy hat, something that he either failed to notice or decided to ignore as he continued to talk about his farm's microscopic denizens.

He told me that some people might be satisfied with just eating organic, pesticide-free food, but that he saw a special advantage to eating foods from soil that he walks through every day. Those bacteria, he said, will be the most beneficial for his intestine. In fact, he appreciated farmers who farmed barefoot, ate their own honey, drank their own milk, and never washed their veggies, since this upped their chances of coming in direct contact with soil microbia. Cody explained that with his new way of farming he would never feed prophylactic antibiotics or dewormers to his cows, a practice that is used in all standard beef operations. Not only do these drugs cost too much and breed bacterial and parasite resistance, but giving a cow these drugs is tantamount to loading them into a crop sprayer and running it directly over the fields. In his holistic system, if one member receives a chemical treatment, its effects will soon be felt by all, including his unpaid workers, the precious soil creatures.[4]

"I want to keep them alive and, yes, I want them in my food," he said.

I thought back to the farm eco-cycle I'd learned about at Jubilee—soil to cow and human and back to soil again—and about the research showing that farm kids in Burkina Faso had healthier gut bacteria than urban kids in

Florence. It seemed that Cody had just handed me another clue to why his holistic system was able to produce such resilient animals.

BAVARIAN STABLES OF PLENTY

Next to "milk" on her flow chart, Erika von Mutius wrote the word "stables."

"By 'stables,' of course, I mean exposure to fungi and bacteria," she said, continuing to enumerate the ways in which traditional farming environments protect kids from asthma and allergy.

She sat down at her computer. For several minutes I watched her click the mouse as she intently searched for something. Then suddenly her expression changed. The stern lines disappeared, replaced by a countenance that I can only describe as merry. She actually gave a soft chuckle. Curious as to what had inspired this change in mood, I walked over to look at her screen. There I saw an unforgettable photo: a rosy infant, fast asleep in a bouncy seat, being sniffed by a very big cow. The bouncy seat had been placed on the stable floor, inches from the cow's hooves, and in the background I could see a woman (presumably the mother) casually loading straw into a wheelbarrow with a pitchfork. A host of other animals looked on.

Erika scrolled through a series of equally striking pictures, all taken during one of her many visits to her farm research sites. They showed toddlers and school-age Bavarian children, dressed in lederhosen, pitching hay, rolling in hay, shoveling manure, or sitting on the backs of their cows. All the kids looked happy . . . and healthy. She pointed at the hay, the animals, and the manure on the stone floors and explained that, from an early age, these children were regularly exposed to the native fungi and bacteria, which helped build a healthy immune system and prevented them from developing asthma and allergies. I thought of Cody wanting to feed his grandkids some manure. Maybe he had a point.

I asked von Mutius how these bacteria and fungi were protecting the Bavarian farm children. She explained that within the intestine and the respiratory tract certain microbes produce metabolites that dampen an allergic response yet promote a healthy immunologic reaction to outside invaders. (In fact, they act much in the same way as the protective lactoferrin proteins in the raw milk that I mentioned earlier in this chapter.) Given what I'd already learned about intestinal bacteria from Justin Sonnenburg, this all made sense: Certain populations can promote health, while others are linked to problems such as colitis and diabetes and, as I was now discovering, asthma and allergies.

But there was one thing that perplexed me. The medical journal article that first led me to Erika von Mutius included a list of microbial species that she and her colleagues found on farms and were presumably involved in creating resilience and decreasing inflammation. Many creatures on the list were types of fungi and molds, such as Aspergillus, that are typically associated with *causing* allergies and respiratory illness. Even more confusing, the bacteria on the list included Listeria and Staphylococcus, and other microbes that are associated with serious infections. These were not at all the types of "healthy" bacteria that manufacturers of probiotics load into their pills, yogurts, and smoothies. In fact, Listeria had recently made a splash in the news after tainted cantaloupe from Colorado was implicated in more than two dozen deaths and several hundred illnesses.

Erika von Mutius agreed. At first glance, it seemed paradoxical that bacteria and fungi typically linked to infections, asthma, and allergies can actually prevent them. But as she and her team began to compare the microbes that surrounded farm children to those encountered by their urban counterparts, this apparent contradiction started to make sense. First, they discovered that the farm children were exposed to a greater diversity of bacterial species. It seems that diversity alone boosts a healthy immune response. Perhaps this is because a mixed population of bugs prevents one nasty organism from setting up shop and colonizing the entire intestine, or perhaps, in concert, these many players create a symphony of messages that keeps the immune system in balance. (Remember, this issue of microbial diversity also emerged in the Burkina Faso–Florence study: The African farm children had a wider array of bacteria than the children in Italy.)

Von Mutius also explained that dangerous-sounding microbes don't all have the same disease-causing potential; in general, those living in the soil produce healthier, more protective metabolites than those in highly populated or enclosed environments. On the wall of her office she had a picture of two Aspergillus molds, one typically grown in a wet, windowless place such as a bathroom and the other found on a farm. Even to my untrained eye, they looked quite different. She mentioned that researchers were trying to isolate the protective substances produced by the farm variety, in order to give them to pregnant mothers and children who are at high risk for asthma, eczema, and allergies.

Some of her colleagues have already begun testing the effects of farm microbes in rodents. They exposed half the mice in their study to two organisms readily found in cowsheds (*Acinetobacter lwoffii* or *Lactococcus lactis*); then they had all the study subjects breathe in a common indoor allergen. The mice that had not been premedicated with the farm bacteria began to sniffle and sneeze, while the others had less of an inflammatory response. Follow-up studies have shown that if the mice are exposed to the indoor allergens and then the farm bacteria, they do not fare as well. This finding suggests that farm exposure starting in infancy (or perhaps pregnancy) is the most protective.

WORMS TOO

After my trip to Munich, I called Cincinnati-based immunologist Marsha Wills-Karp, a member of Erika von Mutius's network of colleagues who study environmental and genetic factors and their connection to childhood asthma. Wills-Karp often collaborates with her husband, Christopher Karp, an infectious and tropical disease specialist, and they share an interest in the link between infection and asthma. I called her to ask specifically about geohelminths, worms that live as happily in the soil as they do in the human or animal gastrointestinal tract. She told me that although some of these worms, like hookworm or ascaris, might cause symptoms when they first enter the gut, they tend to act like many other farm microbes, releasing protective substances and stimulating a counterregulatory, anti-inflammatory response. In other words, the worm and host are involved in a long conversation and are working for mutual benefit, since the worm is able to live indefinitely within the human intestine and the host gets a boost in immunity.

Wills-Karp told me about a study done in a group of Venezuelan children living in a Caracas slum. Initailly, the researchers tested the children's response to airborne allergens such as mold and dust mites and found that they were an impressively unallergic bunch. Then, for the next twenty-two months, they treated half the kids with a monthly dose of a worming agent while the other half were left untreated. When they repeated the allergy tests at the end of the experiment, the treated group had much more of an allergic response than the nontreated one. Wills-Karp explained that all the children involved in the research had probably been primed to react to common allergens, but intestinal worms, acting as counterregulators, had blocked an inflammatory response. When they lost their intestinal parasites, the Venezuelan children also lost their hay fever prevention.

After hearing about the Caracas experiment, the benefits of exposure to animals and farm dirt, and the protective qualities in raw milk, it became clear to me that Cody's assertions had some good scientific backing. Each of these factors played a role in raising resilient animals (and humans). Interestingly, the dirt, the worms, and the farm milk all worked in a similar way: In rare instances they caused illness, but more often they protected against diseases by boosting the host's innate immunity and dampening the host's inflammatory response to allergens and other foreign substances.

But just as I was beginning to think that my visit to Rockin' H amounted to one big lesson in microbiology and immunology, Cody handed me a totally different kind of clue for how to raise healthy children.

SUMO WRESTLERS AND BASKETBALL PLAYERS

Zipping along on his four-wheeler, we cleared a grassy hillock, and suddenly seven hundred rumps came into view—a sea of tawny, black, and white tightly packed into one seven-acre paddock. Cody cut the engine, and we sat and marveled. I could not hear a single sound except for steady munching and the soft plonk of hooves on thick grass. This was bison-style in action.

I helped Cody roll up the wire that separated his beef cows from their fresh paddock and then jumped aside to let them move in. As they trotted by, Cody admitted that converting this herd to an all-grass diet had not been without its challenges; it is one thing to want your animals to behave like wild bison, but it's another thing to get them to do it.

"For the past seventy years or so, we've bred cattle so that they survive on grain. Now I'm telling them that they have to eat grass year-round, even through snow. Many heifers cannot tolerate it. When I started this system, a fair number of my cows just had to go. I lost $80,000 in the first year alone."

Cody explained that finding the right bulls was critical for developing a grass-eating herd. He pointed to one of his favorites, a fellow with a mottled hide and the build of a sumo wrestler—barrel-chested with short, powerful legs. He didn't look a bit like the tall, slab-sided, all-black Anguses that I'd seen grazing along Highway 60 as I drove to Rockin' H.

"If we were to enter a cow from that bull in the county fair, she would be laughed out of the arena," he said, hooting at the very thought of it. "But the judge would not taste her meat or milk." He explained that your ideal grain-fed animal looks like a basketball player, with plenty of room to put on weight quickly, while his own slow-growing grass-eaters look more like this squat bull.

Even after several generations of culling his herd and selecting for grass-eaters, Cody realized that he was not completely satisfied with their performance on his pastures. To be sure, they were healthier than their grain-fed predecessors, but some of them were not as enthusiastic about grazing as he would have liked, and occasionally he still encountered the same respiratory and digestive health problems that had plagued his animals prior to transforming Rockin' H.

At that point, Cody decided that changing the genetics (or "nature") of his cows only got him so far; he also needed to think about the issue of nurture—or what the scientific community now refers to as "epigenetics." These are all the non-DNA factors, including physical environment, diet, and social interactions, that influence which genes will be switched on or off. Cody was especially interested in seeing whether a completely new approach to weaning might improve the health and foraging behaviors of his herd.

CLUE #3: WEANING OFF WEANING

To hear Cody describe it, weaning time on a standard cow/calf ranch is a singularly tragic moment. Most calves are born in the spring or fall and weaned six to eight weeks later. This schedule serves the dual purpose of allowing the baby to beef up on grain so that it can fetch top dollar at auction and letting the mother recover from nursing so that she can easily bear the next season's calf. Weaning entails either completely separating a calf from its mother or, on the more humane ranches, placing the pair on the opposite sides of a fence so that they can still have some nose-to-nose contact.

"There's no time more stressful for that bovine." Cody sighed, referring to both the calf and the mother cow. "There is screaming and hollering for two weeks after the separation. Five to seven percent of calves die. So what do we do? We create a vaccine program and a mineral supplementation program so we feel that we are mitigating that loss. But sometimes it makes it worse. Oh yes, and why don't you castrate them and de-horn them while you're at it?"

After his conversion to bison-style grazing, Cody kept thinking about the wild herds he'd read about in the pioneer's diary. Those bison did just fine without a formal weaning program—their calves grew well, and judging by the enormous size of the herd, the mother cow had no trouble getting pregnant again. Plus, if he kept the herd intact, he would rotate through his paddocks at half the speed and this would allow more time for his soil and grass to recover. So ignoring all conventional wisdom about a nursing calf lowering its mother's fertility or stealing vital nutrients from the next-born, Cody decided to experiment and leave his calves with their mothers and with the herd.

For the first few seasons, Cody was cautious about publicizing his results, aware that he might have to revert back to forced weaning if things did not work out. But by the time I visited, he had enough seasons under his belt to declare his experiment a success and to share the results with his fans in *Acres USA* magazine.

"I don't expect much excitement from mainstream livestock production circles over this new non-weaning program of mine," he wrote. "In fact, I will probably get a lot of ridicule. But that's okay. I'll be the guinea pig!"

Cody discovered that by keeping the calf with its mother, calf mortality rates dropped to near zero. Concerns about the older calf stealing nourishment from its sibling also turned out to unfounded, since a natural weaning usually took place about thirty days before the next calf was born. His "open cow" rate, or infertility rate, was also unaffected; if anything, it went down. And best of all, as these calves grew into young bulls and heifers, they distinguished themselves as healthy eaters: enthusiastic about eating grass and herbs and especially drawn to the most nutritious plants in the field.

"See that young one?" Cody said, pointing to a white cow splattered with black spots who looked like a Jackson Pollock canvas. Despite the dense greenery under-hoof, she was craning her neck under the fence line to nibble at something on the other side.

"She was taught by her mom that she needs a certain forb [grassland species] in her daily diet, and she's getting it over there." He explained that within any paddock on his ranch he could identify more than one hundred forages, each one with a unique taste and a unique nutrient makeup. Dropping to his knees, he began to root through the grass, pointing out all the varieties within our three-foot radius: plantain, lambsquarters (which, he told me, has almost twice the amount of protein as alfalfa), Eastern Grammer Grass, and Buffalo Grass. Cody said that while some grasses were mild-tasting, others presented a more challenging flavor and that his calves needed to gain an appreciation for these, much as a child would for Brussels sprouts.

Cody had a series of explanations for why his nonweaners might have done better on grass than their predecessors. Perhaps the trauma of premature weaning had turned earlier generations on Rockin' H into picky eaters. Or perhaps the sweet supplemental corn-based feed he'd been using for early weaning (which, like standard baby formula, is roughly 40 percent high-fructose corn syrup) had distorted their sense of taste, and made a grass meal seem bitter and unpalatable. He also wondered whether the exposure to formulas and supplements had permanently changed the bacteria in the calf's rumen so that the animal had difficulty digesting grass. But Cody was convinced that the major reason his early weaners did not take well to grass was that they hadn't spent enough time foraging alongside their mothers. He summed it up by saying, "I cannot teach a calf how to graze, but I can offer the mother the opportunity to teach her calf how to graze."

Whatever the reason, by promoting natural weaning and keeping babies with the crowd, Cody had noticed an impressive uptick in enthusiastic grass-eaters and in the wellness of his herd.

NURSING OUR PALATES

After spending time with Cody and his discerning calves, I began to wonder how our earliest life experiences impact our eating choices. I called Julie Mennella, a researcher at the Monell Chemical Senses Center in Philadelphia who has spent much of her career studying factors that influence food preference, especially in children.

Although Mennella wants to better understand the entire "sensory world" of children she's especially focused on the acquisition of bitter and sour tastes, since these flavors are the hallmark of one vital—yet drastically underconsumed—food group: vegetables. Low intake of vegetables in childhood is linked to health problems throughout the life span, including allergy, asthma, heart disease, and diabetes. Mennella explained that children (and presumably calves) are programmed to be a bit more cautious about bitter tastes, since in the natural world this sensation is linked with poisonous or rotten food. But the main reason babies, and then adults, hate vegetables is that they're not exposed early enough. "What we have here is a testament to the power of deprivation," she said. "They never get the experience of eating these foods."

Mennella is a proponent of exposing babies to savory vegetables from the very beginning, and by this she means via the flavorful broth of the amniotic fluid. Her research team identified pregnant women who intended to nurse their babies and divided them into three groups. They asked one group to eat carrots during the third trimester of their pregnancy and then to stop after birth; another was asked to avoid carrots during pregnancy but to eat them while nursing; and the third group was told to avoid carrots altogether. (Carrots were specifically chosen for their distinctive bitter flavor that is readily tasted in amniotic fluid and breast milk.) After weaning, the babies of both groups of carrot-eaters were much more enthusiastic about eating these root vegetables than those whose mothers avoided the food.

"What that carrot study shows," continued Mennella, "is that experiences during pregnancy and the first three months of life [via breast milk] will increase acceptance at weaning, and this may have long-term consequences." As Mennella was telling me about these results, I flashed back to my own daughter, at age four, pestering me for a taste of the breast milk intended for her baby brother. When I finally gave her a teaspoon, she grimaced, complaining that it was too garlicky. At the time, I had been on a real garlic jag, something that is at least partially responsible for her brother's supreme love of garlic.

Another study, done by Mennella's lab, showed that breast-fed babies had a more expansive palate than formula-fed babies, presumably because bitter flavors from the nursing mother's diet showed up in her breast milk. On the other hand, if formula-fed babies were given a bitter, hydrolyzed casein formula such as Nutramigen early on, they were more likely to accept challenging tastes (such as bitter veggies) than those fed the standard sweeter formula.[5]

Julie Mennella explained that while pregnancy and nursing are critical moments for influencing future taste buds, the weaning period is an equally important time for molding a child's food preferences. Once again, if the early foods are bitter and sour, then these tastes are more likely to be appreciated thereafter. Variety and food pairings also make a difference. Much like Cody's cows, who are exposed to dozens of different greens in each paddock, children who are given a greater variety of vegetables tend to eat more vegetables as adults. Combining a challenging vegetable such as broccoli or kale with a sweeter one (squash or carrot) can also boost acceptance. But, cautioned Mennella, taking this approach too far and trying to hide bitter foods by cooking them within sweeter ones might increase intake in the short term but does not help create a long-term vegetable lover.

Mennella posited that giving children farm-fresh (and therefore more flavorful) fruits and vegetables also might increase their likelihood of becoming lifelong fans of these foods. In related research, one survey from the New York region showed that children are more than twice as likely to be avid veggie eaters if most of their meals are prepared at home and if their caregivers shop at a farm stand or farmers' market rather than a supermarket. For school-age children, farm-to-school programs, in which farm-fresh food is served in school lunches, and edible schoolyard curricula that teach students how to grow and prepare their food have also been shown to increase vegetable consumption. (More on this subject in the La Familia Verde chapter.)

Finally, there is something to be said for repeated exposure. Serving challenging foods like Brussels sprouts or mushrooms at different meals and in different combos boosts that food's chance of being accepted. I remember doggedly offering my daughter asparagus on a weekly basis throughout its long California growing season and meeting rejection. And then one day, for some inexplicable reason, she picked up a spear, bit in, and loved it.

During our conversation, Julie Mennella emphasized that the way foods are eaten also has a big influence on lifelong eating patterns. For example, bottle-feeding, even with special "slow" nipples, tends to create faster eaters, perhaps because it is much easier to generate a big stream from a bottle than from a breast; rapid feeding and rapid gulping can then create an enduring pattern of eating faster and more at each meal. (One study showed that infants who sucked more per feeding had a higher likelihood of being overweight at twelve months of age.) Whether breast- or bottle-fed, babies who continue to be offered milk or food after they have turned their head—a subtle cue that they are done—quickly learn to override their inner satiety signals and tend to eat larger-than-average portions, even after weaning. An easy way to avoid this problem is to allow children to feed themselves as soon as they start on solids. Of course, this means that most of the food will end up on the floor, in their hair, and in your hair. Aaah . . . the pleasures of parenting.

Julie Mennella mentioned a number of other environmental factors that can have a lasting impact on novice human eaters, including the general vibe at mealtime. One study reported that children who experience a lot of conflict at mealtime or eat in front of a computer or television are less likely to eat their vegetables, whereas those who have regular family meals tend to have healthier diets. Of course, these issues are difficult to tease out, even in a well-done study. Is it the family meal itself that motivates vegetable eating or are families who eat together and avoid screens at mealtime more likely to eat vegetables? Regardless of the root cause, we can safely conclude that making mealtimes a pleasant, communal experience from day one is a good way to spawn healthy eaters.

Perhaps the most powerful weapon that a parent or caregiver can wield in the "eat-more-veggies" crusade is to simply model good vegetable-eating behavior. Young children are excellent mimics and tend to ape whatever they see us doing, whether it is talking on a cell phone, cursing in a traffic jam, or munching on a stick of celery. I, for one, quickly learned that urging my children to eat challenging food achieved nothing. On the other hand, sitting opposite them and enthusiastically devouring a forkful of kale turned out to be contagious. Apparently this is the case for most children . . . and calves.

"The important lesson in all of this," said Mennella at the end of our conversation, "is that you can't just focus on the kids. The unit is really the family and the community." She acknowledged that one of the hardest things for humans to do is to alter their eating patterns. But she felt that if parents understand how a child's tastes are being influenced from pregnancy onward, it might offer a powerful motivator for making these changes.

After our conversation, I thought about Julie Mennella's final message: Producing healthy eaters is a job for an entire family, an entire community. What she proposed was much more involved than the USDA's "Five a Day" or "Fruits and Veggies: More Matters" campaigns, which simply urge parents to put more produce on their children's plates. She was calling for a multifaceted, holistic approach—just like the one Cody and Dawnnell had instituted for their farm.

ALLERGY-PROOFING FRANKIE

Soon after returning to San Francisco, I saw Frankie, an impish twenty-two-month-old with green eyes and the classic ruddiness around the nose and eyelids of a kid who suffers from chronic allergies. His mother, Susan, brought him into my office for his first visit because he had a cold, and she was worried that he might have another ear infection. She told me that Frankie had been on "too many rounds of antibiotics" in his young life. Just in the past three months he had been diagnosed with two different ear infections and taken three rounds of antibiotics. (The last episode was treated twice since his symptoms did not go away with the first drug.) He also had bouts of an itchy eczema rash on his arms and legs that would improve with topical steroids but then reappear once the creams were discontinued. Because of the allergies, eczema, and ear infections, Susan regarded him as sickly, catching everything that came his way.

As she talked Susan scooped Frankie off my floor, produced a pack of sani-wipes from her bag, and began to mop the snot off her squirming son's face. This prompted him to escape her grasp and dive head-first under my chair. She tried to retrieve him, but when I assured her that he was unlikely to encounter any new or especially dangerous germs down there, she let him be and we continued to chat.

Frankie had been breast-fed for the first three months, but then Susan had to return to work, so he was weaned to a bottle and infant formula. He still absolutely loved his bottle—now filled with organic cow's milk instead of formula—but seemed fairly uninterested in most foods except cheese and white flour pasta. He especially disliked anything that was naturally green (or red, or orange, or blue for that matter), and she felt as if meals were a constant battle.

Susan was frustrated by the frequent colds, the many rounds of antibiotics, the eczema, and the picky eating. She wanted to see if I could offer any advice that was different from what she had heard from Frankie's previous physician. She was also pregnant with her second child and wondered whether anything could be done to ensure that her second child would be healthier. I began to address Susan's concerns and in the course of our conversation mentioned Rockin' H and the Bavarian farm children with their surprisingly low rates of allergy and asthma—and even in some cases eczema and upper respiratory infections. She told me she was intrigued and asked for some farm-derived lessons that she might apply to her city kids. This is what I told her.

1. The Gift of Breast Milk

Susan and I discussed how, by nursing Frankie for those first months of life, she had provided him with a valuable immunologic boost, since breast milk contains disease-fighting factors and also promotes the growth of the infant's protective gut microbiota. These microbiota improve immunity through a variety of mechanisms, many of which are still being elucidated by researchers, including Justin Sonnenburg and Erika von Mutius and the GABRIELA team. Susan looked worried and asked if she had weaned Frankie too early. I explained that while Cody's natural weaning approach is appealing to many human parents as well, there is no evidence that a specific age of weaning is optimal. The general message from organizations such as the American Academy of Pediatrics is that even three months of nursing can offer protective benefits, but that for mother and infant health, more is probably better. Of course, there are mothers who for medical or other reasons cannot nurse at all. Based on available data, it is impossible to say which infant formula is the safest or comes the closest to imitating breast milk, but by following the other recommendations in this list, these infants can also have their best chance to develop a strong immune system.

2. Avoid Unnecessary Inputs

As we learned in the previous chapter, each round of antibiotics could have disrupted the beneficial bug colonies growing in Frankie's intestine, causing a shift to less favorable ones. Research shows that topical steroids can act in a similar way when it comes to disturbing protective microbes and other barriers on our skin. So one big goal for Frankie was to minimize antibiotic and steroid use. To do this we needed a multipronged approach that would include both improving his nutrition and overall resilience and prescribing drugs only when necessary. I explained that there's no question that steroids and antibiotics are life-saving in certain circumstances, but that they are overused. (Americans, on average, receive twenty doses of antibiotics before their eighteenth birthday, and studies estimate that at least 25 percent of these prescriptions are unnecessary.) In most instances, patient education, watchful waiting, and the use of gentler, nonpharmaceutical treatments are more appropriate.

I looked in Frankie's ears and saw that his right eardrum was slightly red. But on the other hand, he had no fever, he looked quite perky, and his runny nose made me suspect that all of these symptoms were caused by a virus, not a bacteria. So, in keeping with published guidelines by the American Academy of Pediatrics, I recommended that Susan forgo antibiotics. Instead, she could give him ear drops made of olive oil infused with mullein, a fuzzy herb with antiviral properties, and garlic. (A study published in the *Journal of Pediatrics* showed that a similar herbal mixture, five drops applied three times daily, worked as well as Amoxicillin for uncomplicated ear infections.) I cautioned Susan that she should not give ear drops if Frankie had discharge or pus coming out of his ear, and we reviewed the signs of an infection that needed more aggressive treatment: an increase in fussiness or lethargy or a fever that lasted more than three days or climbed higher than 102 degrees Fahrenheit.

Frankie had some patches of eczema on his chest, the inner creases of his arms, and behind his knees. They were not angry-looking, and so I encouraged Susan to resist using the steroids, especially since eczema often becomes worse once steroids are discontinued. I suggested that she use calendula flower cream and borage oil instead: both act as natural barriers and anti-inflammatory agents. I also suggested that she replace drying hot showers with bi- to tri-weekly soaks in a warm—not hot—oatmeal bath. (Take a quarter-cup of oatmeal and finely grind it using a Cuisinart, blender, or coffee grinder, dissolve it in two cups of hot water, and then distribute throughout the bathwater.)

Susan raised the issue of immunizations. She had read on a parenting blog that for some children vaccines might increase their susceptibility to allergy and eczema, especially the MMR vaccine. What would I recommend based on my farm experience? I explained that, as part of his holistic approach, Cody had made the decision to no longer vaccinate his herd. He was able to forgo vaccines *because* he had healthier, better nourished, less-stressed cattle, but there was little evidence from the veterinary literature (or the human literature) that his herd was better off because they were not vaccinated. Rather, Cody's calculus was that the dollar cost of vaccines far outweighed their protective value and that a loss of a calf now and then in a herd of seven hundred was an acceptable risk to take. But the cost-benefit analysis looks very different when it comes to human lives. I reassured Susan that large population–based studies and individual clinical trials done in the United States and Europe have shown that, overall, childhood vaccines are safe and effective and that even the most common serious side effects (fever and seizure) are relatively rare and without long-term consequences. Furthermore, analyses of vaccine data have failed to unearth a convincing connection between routine childhood immunization and asthma, allergy, or other forms of immune dysfunction.

But I also believe that, despite all the reassuring research in support of vaccines, it would be scientific hubris to declare that there is no downside to their use or that there is no possible link between vaccines and the worldwide increase in childhood allergic diseases. It is important for researchers to continually monitor each immunization for both efficacy and potential ill effects. One area for further investigation, for example, is identifying the genetic or environmental factors that might cause a small subset of children to fare less well with certain vaccines or vaccine combinations so that they can receive a modified immunization schedule.

3. Farm Therapy

Although formally untested as an intervention, spending time on a sustainable farm offers a low-risk prevention strategy for allergies and asthma. There is some evidence from the Bavarian studies (and now from studies being done with the Amish in Indiana) that exposure to stables and animals while in utero boosts a baby's immunity and lessens allergic or asthmatic reactions after birth. For Frankie, who had already developed allergies, the preventive benefit was less clear, but running around outside, away from the congested city, would undoubtedly do him some good. (When I asked Markus Ege, Erika von Mutius's research colleague, how his research affected his approach to child-rearing, he told me that he now makes more of an effort to bring his young kids to Munich's petting zoo.) Susan had good friends with a small farm outside of Petaluma, California, and thought that this was a good excuse to go there more regularly and help out.

4. Get the Right Foods and Bugs

Next our conversation turned to nutrition. I mentioned what I had learned about farm milk, emphasizing that I didn't think this was a good option unless they were living on a traditional farm and getting the milk fresh from the cow. However, given that the structure of milk is ever more denatured by each phase of processing, I recommended that Susan change to milk that was local (for freshness), organic, not ultrapasteurized, and not homogenized. I also recommended that Frankie drink a little less milk in general, since by filling up on milk he was missing out on valuable vitamins and nutrients found in other foods.

I emphasized that there was no one superfood that Frankie needed, but that good nutrition is based on eating a variety of whole foods, including fruits, vegetables, and grains. The research shows that introducing these foods—or any solid foods—before four to five months of age might increase the risk of allergies and eczema. However, these same studies show that waiting too long after this point in time can have a similarly negative effect. There are multiple explanations for this observation, including the fact that vegetables, legumes, and whole grains are rich in immune-modulating polyphenols and these foods offer children a good source of prebiotic fructo-oligosaccharides, starches that promote the growth of healthy intestinal bacteria.

Speaking of intestinal bacteria, Susan asked about the use of probiotics both for Frankie and for her unborn baby. I told her about recent research showing that when probiotics are prescribed to pregnant and nursing mothers, their children have a lower incidence of asthma, allergy, and eczema. Other studies show that giving probiotics to infants and young children produces similar results. Based on these findings, I recommended a couple of formulations for Susan, Frankie, and baby number two, adding that she could probably get the same benefit from a daily serving of sauerkraut or pickles. Susan laughed, appreciating that this might offer one biological explanation for the stereotypical pickle craving during pregnancy. But then she gave a big sigh. She couldn't get Frankie to eat anything beyond pasta and cheese; pickles and other vegetables seemed out of the question.

5. Raise Adventuresome "Grazers"

At that point I asked Susan about her eating habits and about Frankie's father's as well. They loved pasta, cheese, bread, and chicken. As she ran through her short list, she became visibly flustered. I think it was the first time she'd made the connection between her eating habits and Frankie's, and she resolved to become a better vegetable-munching role model. I agreed that this would be a perfect first step, and I shared the other lessons I'd gathered from Cody and from Julie Mennella: have pleasant family meals, encourage Frankie to feed himself, introduce variety, seek out the most flavorful fruits and vegetables (farm-fresh if available), and repeatedly offer the same food at different mealtimes and in different combinations. I also recommended that all meals take place far away from a computer or television. Then I brought up the carrot study. Susan looked amazed. With each meal choice she made for herself, she was actively shaping the next baby's food preferences! She looked down and gently patted her bulging belly. At that moment, I wondered if she was sharing my vision of a perfect dumpling blissfully afloat in the most flavorful of broths.

ARLEN AT ROCKIN' H

My last day on Rockin' H, Dawnnell mentioned that she and Cody were still short of interns for the upcoming summer, their busiest season. From June through August they needed to raise and slaughter several hundred free-range broiler hens; run their big cow herd; tend to their sheep, goats, and pigs; and grow several acres of vegetables, both for their own consumption and to load into their CSA boxes along with the meat. I couldn't think of anyone to send their way but promised her that I would ask my sixteen-year-old daughter, Arlen, if she had friends who might like to apprentice on a Missouri beef farm.

Arlen mulled it over for ten minutes and then, to my complete amazement, volunteered herself. At that particular time, she was in a phase of high ennui. She had multicolored hair, wore thick black eyeliner, sported many more piercings than your average sixteen-year-old, and was convinced that there was nothing in the world that warranted her interest. Nothing, that is, beyond sleeping, doodling subversive drawings in her oversize sketchbooks, or hanging with a handful of equally tortured Berkeley High students at the Gilman, an all-ages music club best known for spawning the punk band Green Day. Somehow I could not picture her getting up at 6:00 A.M. and doing hard labor on Rockin' H, where the only reprieve would be the Sunday morning service at nearby Hartville Free Will Baptist Church. But now the prospect of beef farming—of all things—had sparked Arlen's interest, and she was not to be deterred.

Two months later, carrying a small knapsack (mostly full of pens and sketchbooks) and wearing her new steel-toed work boots, Arlen boarded a plane to Springfield, Missouri, via Denver. Dawnnell e-mailed that night to tell me that she had arrived. Then, for days, there was silence. This did not alarm me, since Rockin' H is in a valley with very little cell service and only one dial-up Internet connection, but my husband and I were dying to hear how Arlen had taken to farming life. We wondered if she was suffering in silence, counting the days till she could return to Berkeley. Finally, a week into her stay, a photo appeared on my phone, texted by a fellow ranch intern who must have found a pocket of cell coverage. It showed Arlen, a bandana wrapped around her nose and mouth, her eyes shining, holding a kerosene can in one sooty hand and setting fire to a defunct chicken coop. The caption read: "im good."

Another three weeks passed and then, the day before her return, I got this e-mail:

> *Hello Daphne,*
> *Today was Arlen's last day at the ranch and we are going to miss her once she is gone. We enjoyed her and her funny little quirks. She seems to be the kind that can adapt very quickly anywhere she may go. I am certain that if she were to stay much longer the metal rings would come out and the rubber boots would become a permanent fixture, since I fear hillbilly Arlen would eventually take over. She has brightened our day every day. Hope to see you both again soon.*
> *God Bless,*
> *Cody Holmes*

The next day I was standing near the security gate at San Francisco Airport waiting for Arlen to appear. I scanned the stream of passengers, excited to greet her and nervous that I wouldn't spot her in the crowd. Then, sure enough, I almost missed her. Not because I didn't see her, but because I didn't recognize her. The sixteen-year-old I had watched shuffle through the departure gate almost a month earlier had been totally transformed into a smiling, tan woman with popping biceps who strode confidently past me.

It turns out that I had completely underestimated her. She loved everything. The hard work, the long hours, the oppressive heat, the animals, the big farm meals, the raw milk, the other teenage interns who came from rural families and were committed to farming—and most of all, Cody, Dawnnell, and Taylor.

"Being there made me feel important, like I could do anything," Arlen told me as we sat together in the car in the airport parking lot. Then she gave Rockin' H what, for her, is the highest praise imaginable. "It was cool."

In retrospect, Arlen's time with the Holmes family marked a big turning point in her life. She returned to school with a new identity (farm boots and all), more confidence, and a different perspective; many things that had previously held no value for her became interesting. When I think about all the ways in which Cody's holistic system fosters healthy calves and heifers, I have to laugh. Little had I expected when I was on Rockin' H that my own gangly calf would be a beneficiary.

3 HEARTLAND EGG AND ARKANSAS EGG

SUMMERS, ARKANSAS

Pasture-Based Stress Management

The sandpile model (developed by chaos theorists) is an elegant visual metaphor for the cumulative impact of environmental stressors on complex adaptive systems—an impact that is paradoxical by virtue of the fact that the grains of sand being steadily added to the gradually evolving sandpile are the occasion for both its disruption and its repair.

—Martha Stark, "The Sandpile Model: Optimal Stress and Hormesis" (2012)

EVERY SUMMER Gary Cox likes to escape the sticky heat of a northwest Arkansas summer to pan for gold in Alaska.

"I guess you could say that I'm always looking for that golden egg," he mused, tilting casually backwards in a leather office chair.

It was lunchtime and unseasonably hot for April, but Gary still looked fresh, wearing a pressed, button-down shirt, his white hair neatly parted.

In this generic office, with its bare, Formica-topped desk and a fish tank humming in the vestibule, he could easily have been mistaken for a car salesman or an insurance underwriter. The framed painting above his head, a plump hen nuzzling her chick, offered the only clue that Gary was actually an egg farmer.

"A couple of years ago I would have told you that I was very retired and done with the business. You see, I am old school. I started with cages, and I thought I would end with cages. I just wanted to wear out our old equipment and quit."

But then, in 2007, everything changed when Gary and his son and co-owner, Michael Cox, began to bring big changes to their farm. First they converted the existing conventional egg operation to a cage-free, organic one. In the process, they reduced the bird population within each of their five barns from 75,000 to 15,000, and instead of stacking their hens to the ceiling in cages, they set them loose and gave them access to a small outdoor strip of concrete. Next, in 2010, he and Mike decided to start an entirely new business, which they call Heartland Egg. According to Cox, this is the first commercial-scale, low-density, pasture-based egg facility in Arkansas, and it was located just across the dirt road from the office where we were sitting.

Since establishing the Heartland business, things had gotten so interesting that retirement was the last thing on Gary's mind.

"These days, with a more natural system, it all rolls through," he said, pushing an imaginary egg in a perfect circle around his desk. "The birds are less stressed and happier, the eggs are better, I'm happier, and it all makes good economic sense." It seems that Gary has finally discovered his golden egg—actually thousands of them—right in his backyard in Summers, Arkansas.

THE CHICKEN WHISPERER

It was Matt Ohayer, a self-described "chicken whisperer" and the founder of Vital Farms Egg Cooperative, who got me thinking about what an ecological egg production system might teach me about stress. In fact, it was my conversations with Matt that eventually led me to the Coxes in Arkansas. The first time I called him, he was collecting eggs in one of his henhouses. I could hear soft clucking in the background as Matt, cell phone to ear, explained that he had two egg businesses: Onion Creek, a small pasture-based laying hen farm in Austin, Texas, and Vital Farms Egg Cooperative, a virtual farm that served as an innovative packaging and distribution middleman between small- to medium-scale pasture-based egg farms, including Heartland Egg and grocery chains such as Whole Foods. During that first conversation, Matt told me what inspired him to start his business:

"Vital Farms gives farmers a chance to get a higher income per egg laid and to farm in a humane manner, the way that farming used to be. It's a falsehood to call most of the egg farms today farms—they're factories."

After a bit, Matt had to stop talking and go deal with his hens. When I called him back several weeks later to continue our conversation, he was driving his delivery truck up the interstate to visit a Vital Farms member in Oklahoma. I could hear the digital voice of his GPS barking directions in the background. California had just passed a law requiring that all hens be cage-free starting in 2015, and I asked him if there was any objective proof that birds raised in this type of living situation were less stressed and happier, or if this was simply a human sensibility being imposed on hens.

"I can only tell you that my hens are the happiest in the business," he said, referring to the Vital Farms hens, which not only are cage-free but have all-day outdoor access. "However, I don't want to mislead you. From an animal welfare standpoint, you could argue that there still are plenty of stressors." He gave the example of a bobcat that had recently picked off two hundred of his hens over the course of a month.

I asked Matt how often he finds blood spots in his eggs. From my reading, I had learned that these imperfections are a reliable marker of laying hen duress—a blood spot comes from a micro-hemorrhage in the hen's ovary and is a sign that the hen was stressed while the egg was being formed. Matt guessed that his birds produced eggs with blood spots at the same rate as those on any other farm but was quick to add that such imperfections are of no consequence when it comes to egg taste and quality. In fact, they can be a sign of freshness, as they tend to dissipate when the egg sits around for a while.

My conversation with Matt left me confused. Recently, I had started bypassing the $3.50-per-dozen eggs at my local Whole Foods in favor of the $7.90 Vital Farms option. I justified the indulgence by reminding myself that these eggs had a richer, more complex flavor and that it was better to focus on quality over quantity. But in truth it was an investment in the mental and physical health of the chicken herself that ultimately made me feel okay about spending over 100 percent more per dozen. What I had learned about pasture-based farming at Jubilee and Rockin' H had convinced me that this was the healthiest system for animals and for the planet. Now Matt was suggesting that things were not so straightforward: pasture-raised birds were also stressed.

MIKE AND CARL: A TALE OF TWO FLUS

Speaking of stress, there's no ailment that prompts more visits to my medical practice. Sometimes the patient fully acknowledges stress as the primary issue and comes in specifically to discuss relaxation strategies and/or in search of pharmaceutical reprieve, such as an antidepressant, a sleep aid, or an anti-anxiety medication. In other cases, it's a health problem such as irritable bowel, migraine, or low back pain that motivates the visit, with stress as an underlying trigger. But as with chickens, things get confusing with humans, since we all deal with daily stressors, regardless of our environment and our life circumstances. And while some people may crumple under stress, I've seen others who do fine with a fair amount of it. In fact, it's not uncommon for a patient to tell me that stress makes him or her feel healthy and productive.

Two of my patients, Mike and Carl, offer a perfect example of this variability.

Amid all the flu sufferers I treated at the height of the swine flu epidemic in 2009, this pair remains etched in my memory—partly because they paged me within an hour of each other, but mainly because of all the things they had in common. They were both in their midforties, relatively healthy, and married with two kids, and both worked for a big corporation, doing jobs that required a lot of travel. In general, they seemed to face similar life challenges: jet lag, long and erratic hours, time away from the family, work performance evaluations, and a cranky preteen.

But Mike went on to develop pneumonia related to the flu, was hospitalized for three days, and took another two months to recover fully, while Carl got better within a week while lounging in the comfort of his own bed. Why did this happen? Was it just the luck of the draw? Both men were under a fair amount of stress, so what was it about their experiences that explained their very different fates? Strange as it may seem, the Coxes' side-by-side egg operations turned out to be the perfect laboratory for helping me answer this question.

FEATHERS FLYING AT ARKANSAS EGG

Driving west along Arkansas's Route 62, I quickly understood why the University of Arkansas in nearby Fayetteville has one of the biggest poultry research centers in the country. Even the sleepiest of towns in this part of the state has at least one poultry supply store, and every mile or so I passed another egg production compound, looking eerily like a concentration camp with a barbed-wire perimeter and rows of uniform, low-slung rectangular buildings. Occasionally, out in a field, I spotted an old-fashioned laying barn. Abandoned, surrounded by weeds, its wooden overhangs rotting, and torn screens flapping in the wind, it was a relic of a bygone era when laying hens were raised outside, their airy houses used only for roosting and shelter.

Cutting through a corner of Oklahoma, I finally turned onto a bumpy dirt road that led to the Coxes' egg facility. There, inside the reception building, I was greeted by Ashley Swaffar, an earnest young woman with silver hoop earrings, freshly glossed lips, and very clean jeans. Like her boss Gary Cox, Ashley's whole ensemble suggested corporate casual, but she was quick to set me straight when I asked her about her job: "I am first and foremost an egg farmer." She told me that before starting this job, she had been a student in the poultry science program at the University of Arkansas, focusing on poultry reproduction. But midway through her training her hometown friend Mike Cox lured her away from academia, and now she was director of production and quality control at both of his farms: Arkansas Egg and Heartland Egg.

Before starting our tour, Ashley took out her earrings and asked me to remove my jewelry, explaining that jewelry is a bug catcher; as director of quality control, one of her roles is to make sure that things on the farm stay clean. I flashed back to the screaming red BIOSECURE FACILITY signs that had greeted me at the farm's gated entrance. I wasn't sure if these warnings were meant to keep diseases out of the farm or keep them from getting out, but I took off my own earrings, deciding that this question could wait.

Ashley handed me a hairnet (my first of several that day), and we briefly toured the packing plant where white-garbed workers, also wearing hairnets, leaned over rows of sorters and conveyor belts that shuttled eggs in every direction. Then, climbing into Ashley's SUV, we drove less than fifty yards and parked in front of a series of long bunkers with protruding mushroom-shaped vents, identical to the ones I had passed earlier on Highway 62. From these buildings came a continuous high-pitched cackle—the unmistakable sound of a very large, very crowded henhouse. This was Arkansas Egg.

Ashley gave me a fresh hairnet, a baby blue protective suit, and matching booties. The outfit reminded me of a dressing ritual I'd participated in all too often as a medical student and resident: prepping for the operating room. Just then I noticed the butterflies in my chest and my pulse thumping in my ear and had to acknowledge that the prospect of entering a modern henhouse was giving me the same anxious feeling that I would get before entering the high-stakes, hierarchical environment of the OR. I readied myself, took a deep breath, and followed Ashley through the door.

Before my eyes could adjust to the dim light, the piercing smell of ammonia told me I had arrived. I tried mouth-breathing in an attempt to block out the smell, but it was of no use. The stench was too strong, and I found myself wishing I had a face mask to go with the rest of my surgical ensemble.

Just then I noticed that my feet were buzzing. Ashley had warned me about a low-lying electrified wire meant to deter hens from laying their eggs on the unsanitary barn floor, so at first I assumed I was being gently electrocuted. But peering through the haze of dust and flying feathers, I realized that the buzz was from dozens of hens picking at my booties. In fact, the entire floor was carpeted with hens—fifteen thousand of them, to be exact. The peck was not particularly sharp or forceful, and looking closer, I understood why: The tips of their beaks were missing. "They're de-beaked!" I exclaimed over the din. Ashley corrected me. "No, this is not called de-beaking. This is beak trimming, they do it with a hot blade when they are young, and it's for their own good, so they won't peck each other to death." She added: "They don't have that many nerves there."

I shook my bootie at my attackers, but they barely seemed to notice.

Ashley explained that these hens were halfway through a sixty-three-week stay in the house. They first arrived here as seventeen-week-old pullets (brand-new laying hens), and they would stay and lay eggs until they were eighty weeks old, at which point they would be "depopulated." (Depopulated hens become pet food or low-cost packages of mixed white and dark meat for human consumption.) Between each cycle, the henhouses were cleared of feathers and manure. Then they were filled once again with a fresh batch of seventeen-week-old pullets.

I felt the tickle of sweat rolling down my legs and pooling in the elastic ankle-band of my synthetic coverall, and I noticed that Ashley's eyes were watering. It was hard to believe that it could get hotter in here as summer rolled around, or that the intensity of ammonia would double before these hens were done with their laying careers.

As we retreated Ashley pointed to a square of light at the far end of the barn, beyond which lay a dirt and concrete yard about the size of a basketball court. From where we stood, I spotted a couple of hens wandering aimlessly out there. If you wanted a little space, that was definitely the place to go, and yet surprisingly few of the fifteen thousand birds seemed to be taking advantage of the open area.

Outside, as I stripped off my protective gear, I spotted a nearby silo with the word ORGANIC painted on it. That's when it dawned on me that those stacks of egg cartons I had seen earlier in the packaging facility—the ones showing a wholesome lass with bonnet and apron scattering seed for her barnyard hens, the ones that were labeled USDA ORGANIC, FREE-RANGE, FARMERS' MARKET FRESH EGGS—came from this barn! Later that day, in a Walmart closer to Bentonville, I would find this same carton sold for $3.08 a dozen. It was the store's priciest option.

MIKE: IN A SWIRL OF DUST AND FEATHERS

In many respects, Mike reminded me of an Arkansas Egg hen. Whenever I saw him, he seemed caught in a swirl of dust and feathers, and his voice and mannerisms always conveyed a sense of urgency. His was the first of the two phone calls I received that February weekend.

"I think I'm dying," he rasped into the phone. "Please help."

He told me he had a high fever and wracking bone pain. He'd been on the Internet, reading about the deaths connected with the flu epidemic, and was certain that he was its next victim.

Mike never went to the doctor for preventive health visits; he paged me only when something was urgent and could no longer be ignored. Over the years this had included a variety of ailments: low back pain from a slipped disc, a stomach ulcer, a bad case of hives, an inflamed hemorrhoid, an anal fissure (after a particularly long bout of constipation), and a migraine that would not go away with his usual migraine medicine. In the past year his urgent calls seemed to have become more frequent.

Mike worked for a large accounting firm. Once, during one of his urgent visits, he described his typical workweek: "Sunday I take a red-eye to somewhere like Charlotte. I get to the hotel, shower, eat some food, and go work all day, sorting through numbers. Then back to the hotel, sleep for five hours, and fly to somewhere else, say Houston, do the same, and then maybe even a third city. Finally I fly home, where I spend the rest of the week doing the same thing in my own office."

He had little time to exercise and rarely ate anywhere but at his desk during the workday. He felt his job was high-stakes, and yet he had very little control over where he went or what ultimately happened to the information he gathered, since his boss made all the big decisions. At the same time, his work was frequently evaluated, and lately he had sensed that younger and more energetic accountants in the firm were outperforming him. These newcomers did not seem to respect his seniority, and it irked him that they never asked for his advice. On the home front, he said he loved his wife but confessed that he was often short with her. Most days she seemed like just another person who was handing him a to-do list. This, combined with the fact that his once-sweet daughter had become a snarly teen, made Mike almost prefer the road to being at home. When I once asked him if he would consider a career and life change, he laughed. "I'm trapped! I have a big mortgage, a car loan, and lots of other expenses, and there's no way that I can jump into a different lifestyle at forty-three that would make me this much money. In the scheme of things, I'm lucky to have this life."

DUST-BATHING AT HEARTLAND EGG

I shoved my protective wear in the trunk of Ashley's SUV, and we drove out the security gate, past the BIOSECURE warning signs, and crossed a dirt road that separated Arkansas Egg from its younger cousin, Heartland. Suddenly I was reminded of that scene in *The Wizard of Oz* when Dorothy, after being transported by a tornado, opens the door of her drab, black-and-white cabin and steps into the vibrant Technicolor of Oz. Everywhere I looked I saw emerald-green lawns. A gentle breeze cleansed the ammonia from my nostrils. Was it my imagination, or was it actually a little bit sunnier over here?

We pulled up to another white building, newer and taller than the one we had just visited, yet only a quarter of its length. It stood alone at the edge of a large fenced field. Ashley handed me a fresh blue suit, hairnet, and booties, and we repeated the dressing ritual. At this point I couldn't help but feel a twinge of guilt about the pile of nonbiodegradable refuse that my little farm tour was generating. Inside this barn things were cooler than across the road, and while there was still that unmistakable ammonia smell, it was much less intense. I commented on this, and Ashley agreed, offering the caveat that these hens were a few weeks earlier into their stay in the house.

The noise level was also much lower, and we could talk without raising our voices. Perhaps I was reading too much into it, but even the hens' vocalizations sounded less screechy, more clucky . . . almost musical.

There were only five thousand hens in this building, and they all seemed too preoccupied with other activities to bother pecking my feet. For this I was immediately grateful, since I could see that these hens had intact and impressively pointy beaks. Some birds were roosting on what looked like a series of gymnastic parallel bars, others were sitting in a pyramid of laying boxes filled with straw, and many more were playing tag in the clean sawdust that covered the concrete floor. Suddenly a gaggle of hens took flight, soaring up to the highest of the roosting bars. Ashley assured me that the birds in the other house did not have their wings clipped, but I am certain that I saw no flying action there. Perhaps they just had nowhere to go.

Ashley lifted a hangar door connecting the barn to an adjoining three-acre pasture. This door opened daily at about 8:00 A.M. so that all the layers could spend their day outdoors. Immediately, several hundred birds began a casual stroll out onto the grass as we trailed close behind. Outside, some hens started playing tag (or perhaps follow-the-leader), while others engaged in a range of behaviors that I soon learned are unique to chickens. Ashley pointed out the dust-bathers, a subgroup who wriggled like windup toys in small bowls worn into the dirt. Apparently this chicken spa treatment removes itchy mites from the feathers. Other hens were pecking in the ground, rooting for worms and seeds; one particularly industrious one pecked at a patch of gravel. Ashley explained that gravel in the gullet helps break down seed husks and boosts nutrient absorption. At the far end of the yard, under a white sun canopy, a cluster of birds hopped up and down and gossiped breast to breast. Watching them, I felt like a latecomer at an unusually animated cocktail party.

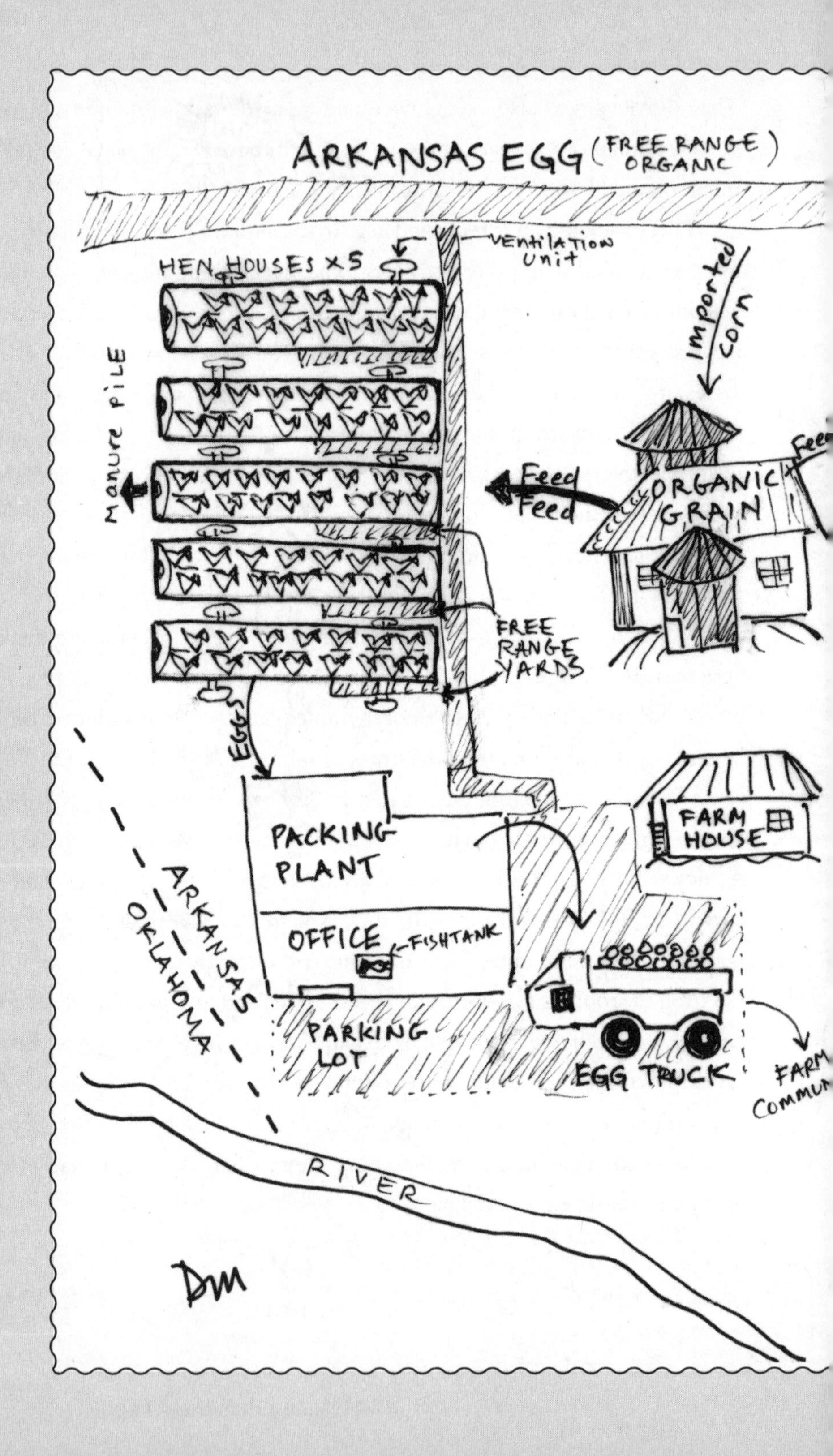
ARKANSAS EGG (FREE RANGE ORGANIC)
HEN HOUSES x5
Ventilation Unit
Imported corn
Manure Pile
Feed
Feed
ORGANIC GRAIN
FREE RANGE YARDS
EGGS
PACKING PLANT
FARM HOUSE
OFFICE
FISHTANK
ARKANSAS
OKLAHOMA
PARKING LOT
EGG TRUCK
RIVER
DM

Arkansas Egg and Heartland Egg

BUSY CARL

If Mike reminded me of an Arkansas Egg hen, then it's fair to say that Carl was more like one from the Heartland operation. His call came in just about an hour after Mike's. He had the exact same symptoms, but sounded surprisingly upbeat for someone with the flu. He assumed he was going to be okay but wanted to find out how he could protect the rest of his family from the illness. He also wanted to know what he might do to speed up his recovery.

Carl, like Mike, worked for a consulting firm. His job was to help companies reinvent themselves. He liked his work because of the team approach his company took and because each assignment posed a unique set of challenges. He also enjoyed the public speaking that was often a part of his job. His work-related travel took him to many of the same cities and office parks that Mike frequented, but Carl was able to plan his trips in advance and set his own schedule. He too was subject to periodic evaluations, although he and his boss had a mutually respectful and supportive relationship and would often run or lift weights together at lunch. He struggled to carve out time to spend with his wife and children, but they enjoyed a family dinner most evenings, and he made it a rule to power down and tune work out when he was at home. His company also gave its employees regular mental health days off from work. In the past, during one of his check-ups, Carl told me that his twelve-year-old son was driving him bonkers but added that he and his wife laughed about it and saw it as a passing phase.

THE STRESS GURU

As I drove away from the Cox farms, I thought about Gary Cox's appraisal of his newest farming endeavor: "The birds are less stressed and happier." I certainly agreed with him. The hens at Heartland appeared to be living in a chicken nirvana, while the Arkansas hens were under extreme duress.

Yet, as Matt Ohayer had cautioned, things were not that simple. In addition to the bobcats, he had listed a surprising number of stressors that are unique to outdoor birds: rainstorms, lightning strikes, hawk and fire ant attacks, parasitic infections, and falls from roosts (which, in a worst-case scenario, can lead to a keel bone fracture and chronic foot pain). And although he felt that hens living the Vital Farms lifestyle were far too content to be aggressive, there was always the risk of being chased and viciously pecked by an intact-beaked alpha chicken. The scientific data seem to support Matt's assertion that pasture-based hens have their share of stress, since studies show that blood levels of cortisol and heterophile antibody, two accepted markers of stress in chickens, are similar in indoor and outdoor birds. (By the way, if you delve into the poultry literature, you will find that advocates of indoor, high-density poultry farming often mention the dangers of outdoor living or cite this equivocal laboratory data to support their perspective.)[1]

So why did Matt and Gary Cox consider the outdoor stressors to be more acceptable than the indoor ones experienced at Arkansas Egg? And why were they convinced that the Heartland hens were happier than their counterparts across the road? Similarly, why did Carl's situation allow him to be healthier and happier than Mike even though they both could be described as having high-pressure lives?

For some answers, I turned to Bruce McEwen, a veteran stress researcher and a senior scientist at Rockefeller University in New York City.

McEwen, who has a Ph.D. in cell biology, told me that his first academic position was at the University of Minnesota, where he studied the cockroach nervous system. In 1966, when Rockefeller University in Manhattan offered him a chance to graduate to vertebrates, he seized the opportunity. One of his first experiments as a young assistant professor involved injecting mice with different radioactively labeled hormones in order to track which parts of the brain were targeted by each substance. He saw that testosterone and estrogen went to the hypothalamus—all fairly predictable. But when he injected the stress hormone cortisol, he was surprised. Instead of simply concentrating in the hypothalamus, a part of the brain already known for its role in the stress response, much of the hormone ended up in a small seahorse-shaped brain structure called the hippocampus. For the next five decades, McEwen's lab focused on understanding the many ways in which the hippocampus, and other parts of the brain that are involved in higher cognitive function, trigger and respond to the experience of stress.

After clearing security at the front gate of Rockefeller University, I found McEwen in a book-filled office on the twelfth floor of an imposing concrete-and-glass research building. Based on his warm greeting and the casual way he directed me to a chair, I guessed that this man suffered little from the condition that had been the focus of his life's work. I told McEwen about my visits to the chicken farms and about Mike and Carl. If he thought it was odd that I was asking about chicken stress and human stress in the same sentence, he didn't show it. Like Justin Sonnenburg, he seemed to be a member of that special tribe of researchers who are quite willing to think about complex problems by making connections across species and across disciplines.

"I am well versed in feather picking in chickens," he said, mentioning that he had co-authored several reviews with poultry researchers in the Netherlands. "In fact," he added, "somc factors connecting stress to disease in humans were first observed in hens."

Stress. I told McEwen that I could think of no other term that is used to describe an emotional state and has such a wide range of applications. The decision whether to buy pastured versus free-range eggs can make us "stressed," and so does living through a war or experiencing a tragic event such as the loss of a loved one. McEwen agreed that the word "stress" has all sorts of connotations, but for most of these experiences we mobilize the same physiologic pathways and the same set of chemical mediators. In some instances this response causes no damage and can perhaps even have a salutary effect, while in other cases it can lead to serious disease.

"Now, let's see," he said, folding his arms across his chest and gazing upward as if landmark studies and their dates were written on the ceiling. "In 1988 Peter Sterling and Joseph Eyer proposed the term 'allostasis' to describe how our cardiovascular system deals with daily stressful events. Later my research collaborative generalized allostasis to include all the adaptive response systems of the body, such as cortisol and the metabolic and immune systems."

McEwen explained that allostasis literally means *achieving stability through change*. Initially, this definition confused me because much of what I have been taught in medicine is based on the principle of homeostasis—the notion that biological parameters, such as oxygen tension, body temperature, and pH, must be kept *unchanged* (or within a narrow window) to keep us alive. But as McEwen proceeded to elaborate on this concept, I understood that while homeostasis applies to these measurements, allostasis describes all the complex systems in our body that vary greatly in their set points in order to maintain overall homeostasis. McEwen gave me the example of cortisol response, which, even in a healthy person, can change radically throughout the day depending on whether the individual is sitting or standing, thinking about a problem, running, or sleeping. Cortisol output is governed by a complex, nonlinear, and dynamic dance between the brain, neurotransmitters, other hormones, muscles, glands, arteries, and nerves.

"Usually allostatic response is turned on or off without leaving a trace," McEwen explained, leading me to imagine this complicated process as being ultimately controlled by a simple light switch. "But if it's used too frequently or not turned on or off efficiently, well, then you have something called 'allostatic overload.' "

In allostatic overload, things go haywire. For example, the hippocampus, a part of the brain that is responsible for memory and learning, shrinks, while the amygdala, a brain zone that reacts to fear, gets bigger; the immune system releases pro-inflammatory cells or cytokines instead of protective fighter white blood cells; and the pancreas overproduces insulin, another hormone of allostasis. Eventually these imbalances contribute to all sorts of health problems: stroke, heart disease, kidney disease, diabetes, arthritis, chronic fatigue, dementia, depression, cancer, infertility, and vulnerability to infection.

"In other words," McEwen said, "the exact same mediators that keep us alive are also able to do us in."

THE PERFECT STRESS LAB

Bruce McEwen smiled when I described the two neighboring facilities in Arkansas with different farming systems but the same management staff, the same supplemental feed, and even the same breed of chickens.

"Aha," he said approvingly. "You have yourself a perfect lab for studying the effects of different kinds of stress." Upon hearing about the thunderstorms, the hawks, the attack of the alpha hen, the falls from the perches, and the parasites, McEwen agreed that Heartland offered plenty of allostatic challenges.

"But," he said, "if you look at these insults in a cluster, you realize that for the most part they are intense yet fleeting. Of course, occasionally they are so severe that they lead to a quick and certain death." In other words, while the daily life of a Heartland hen had its share of heart-pumping experiences, these birds were able to quickly and efficiently mobilize their complex stress response to maintain overall balance. Rarely did they experience allostatic overload.

When I described the housing situation at Arkansas Egg, McEwen nodded knowingly; his past collaboration with poultry researchers had left him all too familiar with these crowded environments. He postulated that instead of the occasional high-drama threat that might occur across the road, the Arkansas hens endured myriad chronic, low-grade stresses that were constantly taxing their adrenals, their brains, their circulatory systems, and all the other components of their stress response.

This most certainly translated into allostatic overload and chronic poor health.

Thinking back on the suffocating minutes I spent in the high-density henhouse, it all made sense. Those hens were protected from hawks and thunderstorms, and there were no perches to present a falling hazard and no soil to harbor ants or parasites. They definitely didn't have enough room to chase each other, and they all had their beaks trimmed so they could never peck each other to death. But at the same time, I could easily generate a list of chronic stressors: sinusitis from poor air quality and high ammonia levels, itching from the mites (which are easily spread in such close quarters and untreatable without a dust bowl bath or a sharp-beaked preen); insomnia from living in a crowd and not being able to retreat to a quiet box or perch; and cramps and osteoporosis from inactivity. And while these hens were beak-trimmed and therefore spared a vicious attack, studies suggest that this procedure, even when done at an early age, can lead to chronic pain and phantom limb syndrome, since the tip of the beak serves a dual function as mouth and arm.

As physically grueling as these conditions might be, McEwen and his colleagues have found that some of the worst stressors produced by overcrowding might fall in the socio-emotional realm. Just as it is common for someone who lives in a city as densely populated as Manhattan to suffer from extreme loneliness, hens experience social isolation when they are too packed in to create a small, cohesive flock. McEwen also mentioned a lack of control as a source of stress. This is what animal welfare experts refer to as loss of "behavioral freedom." When animals of any kind are denied the opportunity to perform their natural behaviors—which for hens includes perching, dust baths, foraging, games of tag, group gossip, and the other activities I witnessed at Heartland—they're much more prone to tip into allostatic overload.

MIKE AND CARL: ON OPPOSITE ENDS OF THE U-CURVE

It was easy to apply what I had learned in Arkansas and at Rockefeller University to Carl and Mike. Carl often faced challenges that might trigger an allostatic response: presenting to a roomful of people, thinking on his feet to offer solutions to a failing company, dealing with his aggravating tween, or running on the treadmill at the gym were all experiences that might cause his muscles to tense, his sphincter to tighten, and his heart to race. But similar to the stresses experienced by the hens at Heartland, Carl's moments of stress were intense yet short-lived. His allostatic response was turned on, but then quickly turned off without leaving a trace.

McEwen sees Carl's daily experiences as falling on the healthy side of what he calls the "U-shaped curve" of stress. Far from being injurious, brief surges in stress hormones can be more beneficial than no stress at all. To illustrate what he meant by a U-curve, McEwen offered me a personal example:

"If I'm about to stand up in front of a crowd and don't have a certain amount of butterflies in my belly, I get worried about the quality of my talk. But then I get equally concerned if I've been talking for twenty minutes and things are still fluttering."

I realized that, unlike Carl, Mike is constantly a-flutter, and he spends his days on the far (unhealthy) end of the U-curve. His stressors are low-grade and chronic: sleep deprivation and jet lag, a sedentary lifestyle and unhealthy food, low back pain from spending too much time at a desk and on a plane, disrespect from his child and his peers, and social isolation. He also is burdened by a sense that he has limited choices and lacks meaningful purpose—feelings, McEwen explained, that are as destructive for humans as they are for chickens. Regardless of the trigger, residing on the far end of the U-curve and constantly activating a fight-or-flight response can lead to allostatic overload and chronic disease. This explains why Mike's immune system was unable to respond appropriately to the swine flu virus and why he—not Carl—ended up in the hospital. It also explains many of his past medical issues, from stomach ulcers to hemorrhoids—all problems that are tempting to treat in a piecemeal, divide-and-conquer fashion rather than by tackling the underlying cause.

THE PRODUCTIVITY PARADOX

When I returned home from Arkansas and began to review my notes, one detail struck me as being at odds with the bigger picture. Both Ashley Swaffar and Robert Higgins, the farm's packaging plant manager, had mentioned that their average indoor, high-density hen laid about 20 percent more eggs per year than the pastured ones at Heartland, and that the Arkansas eggs were considerably larger. Given what I had learned about the U-curve of stress—that episodic, brief stress is more conducive to productivity than chronic stress—this information confused me. How could the Arkansas hens be laying more eggs and bigger ones than their less allostatically challenged cousins? After all, much of the animal welfare literature lists productivity as a key indicator of animal wellness.

I could not dismiss this difference in output as simply a result of bird breeding since, as I mentioned, the Coxes were using the same Cinterion breed of hen (a cross between a White Leghorn and a Rhode Island Red) in both setups. It also could not be explained by type of feed, because all the hens on their property were receiving identical organic mash. Also, I soon realized that these findings were not isolated to the Coxes' facilities but were in keeping with national trends. In the midtwentieth century, most laying hens in the United States were pastured on small farms and each hen produced, on average, one hundred eggs per year. By contrast, today's standard battery-caged hens, concentrated in far fewer and larger facilities, lay about three hundred eggs per year, with egg size at least one-third larger than it was in the 1950s.

I thought about Gary Cox rolling his imaginary golden egg around his desk. Here was a bottom-line guy, a serious businessman, who was truly enamored of his pastured Heartland chickens despite their lower laying rates. I decided that I must be missing something. Could it be that focusing on egg size and number of eggs laid per hen was the wrong way to measure productivity? I needed to do some more investigating.

I called Matt Ohayer, once again catching him in one of his henhouses. In the background I could hear sounds that I now recognized as the satisfied clucks of pastured hens. Matt's explanation for this productivity paradox was twofold. First and foremost, he explained, it had to do with boredom.

"Those Arkansas hens have nothing better to do than eat all day," he said. I remembered how they converged on my booties, and I imagined that when a visitor was not in their midst they would channel all that pecking into eating grain. They were doing exactly what we tend to do when we're anxious, bored, and inactive: self-soothe by grazing in the fridge. The pastured hens, on the other hand, had plenty of distractions. They could get so busy chasing, foraging, dust-bathing, and roosting that they might actually skip a meal, and they'd certainly burn more calories, regardless of how much they ate. Pregnant women who eat more daily calories and exercise less tend to give birth to bigger babies, and it made sense that the same would hold true for chickens and their eggs.

Matt also explained that the controlled environment of a barn allowed farmers to trick hens into hyperproductivity. In a pastured system, a hen's laying pattern is tied to the cycles of the sun and the seasons, but when she's largely confined indoors, such as at Arkansas Egg, it is possible to manipulate her exposure to light and induce her to lay sooner and more often.

Looking at the bigger picture, I started to see that those Arkansas hens were not as productive as I initially thought they were. First of all, they subsisted only on corn mash, whereas the Heartlands got a share of their calories from grubs, seeds, and anything else they found in their pastures. As fuel and corn prices continue to escalate, the cost that goes into making each indoor egg will also start to climb. Given that these Walmart-bound eggs fetch less than half the price of the Heartland ones, you start to see why Gary Cox is so enthusiastic about his new farming setup.

Then, if you consider the nutritional value of each egg, the equation really shifts. The crowded, indoor layers are definitely sacrificing quality for quantity. Studies show that sedentary hens that eat only mash produce a less nutritious egg than hens that forage outdoors and eat grass and worms. Specific nutrients mentioned in these studies include vitamins D, A, and E and omega-3 fats. Taste is another important factor. While there are no published trials examining whether chronic stress delivers less palatable eggs, related research shows that chronically stressed animals, when slaughtered, produce meat with an inferior taste, texture, and nutrient makeup. One can guess that the same might hold true for eggs. (In informal taste tests in my kitchen, pastured eggs always emerge the winner.)

Shell quality is also a consideration. Hens that forage tend to take in more vitamin D and methionine (from sunshine and worms, respectively), and this translates into a hardier shell. So while Arkansas hens may lay more eggs, the eggs laid at Heartland have a better chance of making it to the supermarket shelf. When I toured the Coxes' egg-packing facility, I asked the plant manager, Robert Higgins, why the floor was so strewn with broken eggs. Above the clang of the sorting and packing machines, he told me it was an Arkansas Egg day. "I always know when I'm packing Heartland," he said, "because there are fewer broken eggs in here."

Then there was the issue of cloacal prolapse, the cloaca being the multipurpose opening that serves as a combination anus, urethra, and vagina in hens. Pushing out a large egg puts a hen at greater risk for developing something that, to my eye, looks like a massive hemorrhoid. But it gets worse—apparently it is a universal chicken instinct to peck at vibrant colors, and a bright-red, bulging cloaca is no exception. You can picture the rest. . . .

Farm worker health is another factor to plug into the productivity equation. From those few stifling minutes in the Arkansas Egg henhouse, I can easily believe the studies reporting that poultry workers who spend their days in these high-density environments suffer from more lower and upper respiratory symptoms and lower lung function than people in almost any other occupation. In the short term, daily exposure to mites, dust, endotoxins, and gasses causes more asthma and wheezing, and in the long term it leads to disabling lung diseases such as emphysema. While these costs are unacceptably high for the workers themselves, farm owners like the Coxes bear the burden as well with higher health insurance and worker's compensation insurance premiums, higher rates of worker disability, and higher employee turnover.

The more I researched, the more I uncovered hidden costs associated with those extra (and extra-large) eggs at Arkansas Egg. It made me realize that productivity—at least the definition used by most hen producers—is a lousy metric for animal health, since it focuses on short-term gains rather than long-term wellness. In the end, this shortsightedness does not serve the best interests of the farmer, the farm, the farm worker, the consumer, or—of course—the chicken.

Interestingly enough, I often see this same narrow view taken in regard to human productivity. So many of my patients work for companies where the work ethic is one of intense competition, with a standard sixty- to eighty-hour workweek—all in the name of greater annual profits. These patients tell me that they rarely take a vacation and wouldn't dream of exercising during their lunch break as that would be interpreted as a lack of dedication to their job. Patients like Mike adopt these values and desperately try to mortgage their health in order to maximize their work performance and their revenue; they pull all-nighters, skip meals, eat food that offers fuel but few nutrients, use caffeine to stay awake and then alcohol to go to sleep, and rarely move their bodies except to hop on a plane or walk from desk to car. But this approach is not sustainable and eventually leads to all the chronic health problems associated with allostatic overload, including high blood pressure and blood sugar, weight gain, and sleep apnea. More often than not, they're also overwhelmed by depression or anxiety.

Not surprisingly, newer research is showing that this workaholic approach is unhealthy for businesses as much as it is for individuals. Maximally taxing employees translates into lower work performance and, in the long term, less financial success. These findings have prompted forward-thinking companies, like Google, Facebook, Pixar, Zappos, and Patagonia, to offer their employees more time for relaxation, exercise, social connection, and pursuit of outside interests, as well as a generous amount of sick leave.

PROMOTING PLASTICITY

If you ask Bruce McEwen whether there's hope for someone like Mike, his answer is surprisingly rosy: His studies have shown that when the stress response abates, parts of the body show signs of recovery. "Of course, this doesn't mean that plaque will regress from the arteries," he adds. "But we've looked at the hippocampus's ability to respond to short-, medium-, and long-term stress, and those neurons are forgiving. They can shrink and then recover. Even with extreme stress and depression, those neurons just shrink, they don't die."

McEwen referred to this ability to recover as "plasticity," explaining that some humans (and chickens) are innately more plastic and that, to a certain degree, plasticity is determined by genetics.[2] But much can also be attributed to epigenetics, those influences that aren't found in the DNA—things such as how one was treated in infancy and maybe even in utero. For example, rodents that received more maternal licks as pups show more attachment, less stress and emotional reactivity, and slower cognitive decline and have a longer life span than those who were not licked. Researchers are now finding the same thing to be the case with humans: Early life traumas can have a negative impact on coping skills throughout the life span, while nurturing in infancy can create real plasticity, even in the face of severe stressors down the road. Despite the critical impact of those early years, McEwen believes that plasticity can be improved at any phase of life by making specific behavioral and lifestyle choices.

"You know what I think is really neat?" he said, sounding as if he were going to share some handy new trick. "You can take people in their fifties and sixties who are depressed and have them exercise for an hour a day. And guess what happens? They have more blood flow in their prefrontal cortex and their memory gets much better. And their hippocampus [the memory center in the brain] gets larger as measured by structural MRI."

"And you know what's even neater?" he added, leaning forward in his chair. "You can take adults with chronic anxiety disorder and expose them to a mindfulness program. And they have a shrinkage in their amygdala [the fear center in the brain]."

I asked Bruce McEwen what he saw as the role of antidepressants in coping with stress and maintaining allostasis. He explained that while research shows that these drugs can facilitate plasticity, they really only work in tandem with positive lifestyle changes, such as behavioral therapy, meditation, increased exercise, dietary changes, and what he called "a positive environment."

"In a negative environment, antidepressants are not helpful." Then he added: "They may even make things worse and push someone to suicide."

POULTRY PLASTICITY

In my final conversation with Matt Ohayer, I realized that chickens can also exhibit plasticity. He told me that when he mentors farmers who are trying to start a pasture-based egg business, he often suggests they get "spent hens" to use as "training wheels." Spent hens are birds that have completed their sixty-three weeks in the poultry barn; since they are no longer considered suitably productive, their next stop is typically pet chow. When these retired hens are first transferred to a pasture setting like Heartland, they usually hide in a corner and look like any other stressed-out battery chicken. But soon they start to venture outside to join their cavorting sisters.

"With a little sunshine, outdoor play, and socializing, these hens can get back up to 80 percent productivity for at least another thirty weeks," said Matt. I pictured a dancing hen, her hippocampus swelling like a ripe plum while her amygdala shriveled down to nothing.

Given that early life experiences can have a lifelong impact on reactions to stress, I asked Matt where he got his pasture-based birds.

"We have our hands full with the layers and wouldn't like to take control of the breeding," he answered. He gets his newly hatched hens from Hy-Line, one of two major hatcheries in the United States that supply more than 90 percent of the chicks to egg farms. (Compare this with the 1930s, when there were more than thirteen thousand hatcheries in the United States.) In other words, the same sources of chicks and the same breeds are used in the huge indoor hen operations *and* in many smaller pasture-based farms. Additionally, most egg farmers—even those who are pasture-based—keep their chicks in enclosed barns for the first seventeen weeks of life and turn them out to pasture only once they become layers. I told Matt about Frank Reese, a Kansas heritage bird farmer with whom I had spoken several weeks earlier. Reese hatched and raised his own laying hens, getting them on pasture from day one. He also raised heritage turkeys and other meat poultry and saw his farm as truly sustainable because he bred his own birds and didn't rely on outside inputs.

"I think it's great that there are farmers who can do that," Matt replied. "But they can only feed a few people. It's a big world."

I hung up from this last conversation feeling a little disheartened. Matt and his colleagues were doing a far better job than most, and his eggs are delicious. Still, I couldn't help but think that for $7.90 a dozen it would be nice to know that the hen in question had been pastured for her entire life and not just at the stage when she began to produce her eggs. But the reality was that massive operations like Hy-Line and Centurion could breed and hatch chicks much less expensively than Vital Farms ever could. And if we, the consumers, are to ever start demanding that our pasture-based eggs be laid by hens who themselves were born in a pasture, those $7.90 cartons will need to be marked up considerably more. How much are we willing to pay to ensure that our laying hens are engaging with nature throughout their lives?

Thinking about all this, I noticed that I felt stressed.

THE STRESS REDUCTION TOOLBOX, OR, HOW TO BECOME A PASTURE-BASED HUMAN

Soon after visiting the Coxes' farms, I called Alice Domar, a stress researcher at Harvard. She has authored many articles and books on stress management and has also founded a center for mind-body health in Waltham, Massachusetts. I was impressed by what Bruce McEwen had told me about brain plasticity and wanted to get Domar's advice about specific activities that could reduce chronic stress, restore allostasis, and foster plasticity. She had recently completed a study showing that mind-body support groups help women with infertility become pregnant.

"You can't just tell people to relax," said Domar. "Clearly that's not going to work."

I asked her about courses that teach relaxation and breathing techniques, and she sounded fairly dubious about those as well.

"For most people," she explained, "their anxiety level is so high that they can't get much out of sitting in a room and breathing deeply. You have to give them a whole toolbox of skills to pick out what will work." Domar believed that a complex, multipronged problem like stress demands a complex, multipronged solution.

Unpacking Domar's toolbox, everything looked familiar. These were skills and behaviors that I had noticed in Carl and the Heartland chickens. They were also practically identical to the ones Bruce McEwen had mentioned when I asked him how to re-expand those hippocampus neurons and shrink the ones in the amygdala.

TOOL #1: JOIN THE FLOCK

Neuroscientists at the University of Chicago have shown that loneliness raises cortisol, lowers immune function, and causes all the other negative physiologic effects that we discussed earlier in this chapter. So how does one counteract loneliness? Interestingly enough, it doesn't matter how many people you cross paths with on a daily basis but rather how intensely you feel connected to them. I have some patients who seem to live fairly solitary lives yet feel extremely tied to a community, while others, like Mike, have colleagues and a family but feel isolated. Virtual (Internet) communication can enhance personal connection but is generally not a substitute and can sometimes increase isolation.

Here are some antidotes to loneliness:

- Find a club that focuses on one of your passions. (For outdoor enthusiasts, the Sierra Club is one of my personal favorites. For the crafty DIY types, your local cooperative studios or tech shops are a great option.)
- Join a sports team or train with others for a fund-raiser such as the Arthritis Walk, the AIDS Walk, the Breast Cancer Walk, or Team in Training.
- Meet with a religious or spiritual group.
- Subscribe to a political or social cause.
- Join a support group that will help you make positive changes to deal with a chronic health issue. This could be anything from AA to an arthritis self-management program to Weight Watchers. Check online for a local group that matches your interests.
- Volunteer for a local organization such as a school, library, public park, elder care center, pet shelter, community clinic, or food bank. VolunteerMatch.com is a national website that will give you a list of local volunteer opportunities organized by area of interest.
- Take time in your day to talk to someone about interests or experiences that you both have in common. It can be something as mundane as a book, a sport, or a movie. Even small positive connections throughout the day will make you feel better about yourself and others.

If you feel extra challenged when it comes to forging positive relationships or always feel that you have negative interactions with others, I would recommend a social-emotional learning program. These programs give you the skills to better communicate with others. For this, I especially admire Marshall Rosenberg's Non-Violent Communication Trainings, which are offered worldwide (www.cnvc.org), and for couples I recommend Sue Johnson's Emotionally Focused Therapy (www.iceeft.com).

TOOL #2: ENHANCE YOUR BEHAVIORAL FREEDOM

There are thousands of self-help books and online tools to help you take control of your life and increase your sense of purpose. In an ideal world, everyone would avoid jobs and relationships that make them feel powerless or unimportant. However, even in the most toxic environment there are steps you can take to optimize your situation: Set realistic goals for yourself and communicate them to others, exercise daily, and take time to care for and pamper yourself.

CBT, or cognitive behavioral therapy, is a short-term, focused therapeutic intervention that helps you understand the direct connections between your thoughts, feelings, and certain behaviors. It has been shown to help quell anxiety and improve one's sense of control. When selecting a therapist, make sure he or she has been formally trained in this technique.

Ancient contemplative techniques such as meditation and yoga have also been shown to increase self-esteem and reverse the chronic negative effects of stress. In one of his articles, McEwen explains that these pursuits cultivate compassion and kindness, two positive emotions that seem to play a key role in shrinking the amygdala (the anxiety center) and increasing the size of the hippocampus. Not surprisingly, studies show that experienced practitioners of mindfulness show more plasticity than those who occasionally say a couple "ohms."

TOOL #3: PLAY TAG AND DUST-BATHE OFTEN

Bruce McEwen told me that exercise rejuvenates neurons, especially in the brain's hippocampus, and improves nerve cell survival. But here's the catch: this has been shown to occur only when the exercise is voluntary and enjoyable. If one is forced to exercise, the brain doesn't show these positive changes. As an added benefit, maintaining a healthy weight and good cardiovascular fitness through daily exercise boosts self-esteem and therefore decreases your level of stress.

TOOL #4: GET A GOOD NIGHT'S ROOST

Studies show that sleep deprivation produces the same effects in the brain as chronic anxiety and depression. (Of course, this can be challenging to tease out, because all three factors are interrelated.) Nonetheless, a good night's sleep is absolutely critical for counteracting stress. Make sure you have a sleep routine and a safe, quiet place to sleep. To avoid the 3:00 A.M. awakenings, exercise daily, avoid alcohol and caffeine, don't eat too close to bedtime, and make sure you're not taking a medication that might be disrupting your sleep. Most important, don't get stressed out worrying about the length of your sleep! Studies show that some people do just fine with six hours of shut-eye while others need much more. If sleep is an issue, consult a health care provider.

TOOL #5: PUT THE RIGHT FOODS IN YOUR GIZZARD

Stress, via hormones and neurotransmitters, can affect our eating habits. When stressed, we go for high-fat, high-sugar, and overall high-calorie foods—the junkier the better. Recent research also shows that when we eat these foods, soothing substances such as endocannabinoids are released from our gut, completing the feedback loop and pushing us to make these same selections whenever we feel tense or anxious. Clearly this instinct meets a short-term need at the expense of our long-term health, since the additional calories (and the oxidative effects of these sugars and fats) contribute to higher insulin levels, more inflammation, and all the chronic health problems associated with stress. Besides just saying no to food that is excessively cheesy, sweet, salty, or highly processed, are there things we can eat that are calming and good for us? Better yet, are there foods that can prevent us from getting stressed out in the first place? Here the research gets rather sketchy, since there are so many variables at play that it's hard to pinpoint the possible role of specific foods or food combinations in stress reduction. Therefore, understand that what I suggest makes good sense based on what we know, but that it has not yet been proven through well-designed research studies.

Four amino acids—tryptophan, phenylalanine, l-glutamine, and tyrosine—are big players in helping our brains produce stress-proofing, mood-enhancing neurotransmitters such as serotonin, norepinephrine, epinephrine, and dopamine, and foods rich in these amino acids might increase blood and brain levels of these important building blocks. Interestingly enough, eggs (specifically egg whites) and chicken are good sources for all four amino acids, as are foods preferentially eaten by hens: seeds, nuts, and whole grains. And yes, those proteinaceous worms and bugs also offer a ready supply of these four amino acids.

Foods rich in antioxidants and omega-3 fats help with nerve function. They also counteract inflammation, helping to prevent the secondary diseases caused by stress. The greens foraged by chickens are excellent sources for both of these nutrients, and leafy greens are a good choice for humans as well. Spices and herbs of all varieties are extra powerful sources of antioxidants. Pastured eggs themselves happen to be a decent source of antioxidants and healthy fats, with one study showing that they contain 66 percent more vitamin A, 300 percent more vitamin E, 700 percent more beta carotene, and 200 percent more omega-3 fats than your standard caged egg. Vitamin D deficiency is associated with a variety of chronic health problems, including diabetes, heart disease, and depression. While it is hard to get this nutrient from diet alone, pastured eggs (specifically the yolks) are one of your best sources of vitamin D, delivering approximately 40 percent of the RDA per egg.

Speaking of herbs, Gerry Huff, a poultry researcher at the University of Arkansas in Fayetteville, told me that her lab was investigating an Indian herbal combination called Stresroak that seemed to help reduce stress in her birds. She was kind enough to e-mail me the list of ingredients and I was intrigued when I recognized three of them as ones that I use to alleviate stress in my patients. They're listed in the table below with the recommended preparations and doses for adults, developed in consultation with my favorite herbal medicine professor, Tieraona Low Dog, M.D. These are generally safe formulations, but I would take them only with the approval of your health care provider.

HERB NAME	DOSE	PREPARATION
Ocimum sanctum (Holy basil, Tulsi)	800–1,200 mg/day taken in 3 divided doses	Standardized to 2.5% ursolic acid
Withania somnifera (Aswaganda)	1–6 g/day of the crude root or 1,000–1,500 mg/day of extract	Standardized to 2.5% withanolides
Rhodiola rosea (Golden root)	100–500 mg/day taken in 2–3 divided doses	Standardized to 3–6% total rosavins and minimum of 0.8% salidroside

POSTSCRIPT ON MIKE . . . AND ON PRODUCTIVITY

Mike continued to be an allostatically overloaded mess for several years after his recovery from swine flu. He would appear in my waiting room from time to time with stress-related ailments, from reflux to psoriasis to hemorrhoids, but seemed unwilling or unable to focus on the root causes of these problems. Then, one recent spring day, he surprised me by showing up for his first-ever prevention visit. He told me that things had gotten so stressful that he had finally met with his boss and requested, in exchange for a small cut in benefits, less travel and one personal day every two weeks. To Mike's surprise, his employer was receptive to this request and even seemed to respect him for making it. Interestingly enough, these changes were a catalyst for more new things: Mike started to exercise, to eat at home more often, and to bring homemade lunches to work. Soon he noticed that his hemorrhoids and skin were getting better and that his sleep patterns were healthier. And when he got on the scale in my office, he was delighted to discover that he had lost ten pounds. "Funny," he said, "I'm working fewer hours, making less money, and no longer killing myself—but every day I feel so much more productive."

Hearing Mike, I remembered a conversation I had with Anne Fanatico, a professor of sustainable farming and a pasture-based poultry expert at Appalachian State University. She's also a friend of Ashley Swaffar's. I asked Fanatico how she defined productivity.

"We know that number of eggs laid cannot be the only measure," she said. "In fact, animal welfare people say that if a hen is laying more than 250 eggs per year, it will be bad for her." Fanatico then mentioned many of the factors that I listed earlier as important things to consider when discussing productivity: worker's health, egg quality, energy inputs, and so on.

"But is anyone working to redefine productivity?" I asked.

"Why, yes," she said, giving a patient laugh. "This is really what the whole sustainable agricultural movement is about. People have different ways of describing it. Some call it the triple bottom line, or the three E's (equity, economy, and environment), or the three P's (people, planet, production), but in the end we're looking at a long list of interrelated factors in order to describe sustainability."

Once again, I was just discovering new ways to assess the health of a complex system, while eco-agriculturists like Anne Fanatico had already given such ideas years of consideration.

Integrated Pest Management as a New Approach to Cancer Care

The time has come in America when the same kind of concentrated effort that split the atom and took man to the moon should be turned toward conquering [cancer].

—President Richard Nixon,
State of the Union Address, January 1971

This is the environmental awakening. . . . Partnership with nature replaces cavalier assumptions that we can play God with our surroundings and survive. It is leading to broad reforms in action, as [we] mobilize to conserve resources, to control pollution, to anticipate and prevent emerging environmental problems, to manage the land more wisely, and to preserve wildness.

—President Richard Nixon,
Special Message to Congress, February 1972

AS A consequence of spending time on farms, I began to notice just how often my patients use agricultural metaphors to discuss their health problems. For example, minutes after I met Dava, she asked:

"Can't we just yank it out by the roots or kill it with some kind of weed poison?"

She was referring to precancerous cells that had recently been discovered in her esophagus.

A compact, high-energy woman in her midfifties, Dava was used to taking care of business and not letting problems simmer. But she had suffered from GERD (gastroesophageal reflux) for years despite a concerted effort to squelch the daily symptoms. She tried eliminating a whole list of foods, including coffee, chocolate, spicy food (her favorite), and alcohol; she raised the head of her bed; she'd taken countless rounds of Tums; and eventually she turned to stronger acid blockers, including the proton pump inhibitors (PPIs). Each medication had some benefit, although her symptoms never completely disappeared.

Finally she consulted a gastroenterologist who, as part of her work-up, performed an esophagastroduodenoscopy (EGD), a procedure in which he inserted an endoscope through her mouth in order to examine (and biopsy) her lower esophagus. Dava first came to see me carrying this troubling EGD report, which said that a segment of cells lining her lower esophagus had morphed in both structure and function; instead of the pale, glossy squamous cells that normally inhabit this region, hers were coarse and beefy red, resembling those typically found in the stomach and small intestine. This transformation in both DNA and cell structure is found in about 1 to 6 percent of the general population and is called Barrett's esophagus, after the Australian surgeon who first described the phenomenon. Barrett's is likely to be the way the esophagus deals with years of exposure to misdirected stomach acid and is considered a precancerous state because each year about one in two hundred patients with this diagnosis go on to develop adenocarcinoma of the esophagus—the most common form of esophageal cancer. And while esophageal adenocarcinoma is still relatively rare, since the 1990s its incidence has risen at a faster rate than almost any other cancer. Experts cite our Western lifestyle, including diet and environmental exposures, as a likely explanation for this statistic.

Dava told me that she considers herself a calm and methodical person, but the word "cancer," even with a big "pre-" in front of it, sent her into a panic. What alarmed her most was that her gastroenterologist told her there was no proven treatment for reversing the Barrett's esophagus process. Even the medication she had been prescribed to block the release of stomach acid had not been shown to make a significant difference. All he could offer was more intensive surveillance—in other words, frequent EGDs with biopsies to ensure that she was not in the unfortunate minority who developed cancer. This proposal for no action and long-term watchful waiting using a test that was uncomfortable, expensive, and not at all risk-free left Dava feeling "freaked out," as she told me. She felt as if her throat was inhabited by some kind of weed or alien pest—hence her wish for total eradication.

BEYOND WHACK-A-MOLE CANCER TREATMENT

Within the context of modern cancer care, Dava's demand was quite reasonable. In fact, "cut it out" or "poison it" pretty much captures the standard approach used by oncologists to treat both early and advanced cancer. The esophagus, however, does not fare well with these standard therapies. A paper-thin, narrow tube squashed behind the trachea and heart and in front of the spinal column, it is easily prone to perforation and strictures, and any surgeon will tell you that few operations carry a higher mortality rate than the removal of all or part of the esophagus. Naturally Dava felt singularly unlucky that her precancer was in one of the few organs that could not withstand the usual armamentarium of cancer-fighting therapies.

But Brian Reid, gastroenterologist and veteran Barrett's and esophageal cancer researcher at Seattle's Fred Hutchinson Cancer Research Center, would argue otherwise. He thinks that the unforgiving nature of the esophagus forces researchers, physicians, and the general public to face the ugly truth about cancer in a way that more resilient organs (such as the lung, colon, or breast) allow them to ignore—namely, that most of our eradication efforts are not working. I called him one summer day to talk to him about his work and was lucky to catch him between his clinic and his lab.

"We can do the surgeries and blast cancers with the chemotherapy, but overall we've had forty years of frustration in treating this disease. It's like the 'whack-a-mole' game—you hit it down in one place in the body and eventually it pops up in another." Despite billions of dollars of research, mortality rates from most types of malignancies have not budged (and some have even increased) since 1971, the year President Richard Nixon signed the National Cancer Act and called for a "war on cancer."

The reason for these sobering statistics, Reid explained, is that scientists and the general public misunderstand the disease.

"What's clear to me now," he said, "is that every step of cancer is a dynamic and unpredictable evolutionary process with multilevel interactions between mutated cells and the surrounding normal tissue. And yet the vast majority of the cancer world treats it as a relatively static occurrence: one kind of mutated cell run amok that never changes and only divides. You see, it's easier to assume that it is static. The fact that it isn't and that people are still dying at the same rate as before—well, that's just collateral damage."

Reid then offered me three reasons why he takes issue with the standard approach to cancer care. First, chemotherapy can promote the growth of resistant cells. Referring to Darwin's theory of natural selection, he explained that vulnerable cells die off with treatment, leaving room for hardier ones to replicate and repopulate with a vengeance. Eventually it is these surviving cells that spread, forming deadly metastases in distant organs. Second, the standard genetic analysis used to determine the severity or treatment for a given cancer fails to consider that malignant cells are constantly interacting with the surrounding tissues. In other words, two cancer cells with a near-identical DNA profile can behave differently depending on local environmental factors such as pH or oxygen and glucose concentrations. Finally, cutting out the cancer poses its own set of problems since it's easy for a surgeon to leave behind a few malignant cells lying at the margin of the main tumor or to overlook early microscopic metastases known commonly in the field as "micromets."

"So you see," Reid said, "especially when it comes to aggressive or metastatic cancer, we really need to start doing things differently." He swiftly followed this remark with the caveat that his work on alternative approaches is still experimental; until more is known, he was not suggesting that people with cancer abandon or deviate from their current therapies.

QUADRATIC EQUATIONS AND CABBAGE MOTHS

Reid is not the only scientist who's challenging the predominant ways in which medicine understands and treats cancer. His colleagues include a small but growing group of scientists, scattered around the globe, who refer to themselves as integrative evolutionary cancer researchers. More than two hundred of these researchers recently attended a meeting, an impressive number given that their first convening in 2007 was so small that, according to Reid, "a small fire could've wiped us out." These "evo" cancer researchers are an eclectic bunch—astrophysicists, mathematicians, evolutionary biologists, ecologists, psychologists, and even a few oncologists, all united in a quest to develop new models to better understand how cancers develop and spread.

But within this adventuresome group, radiologist and mathematical oncologist Bob Gatenby stands out for having traveled further afield than most. He has turned to agriculture for new ideas on how to approach cancer care.

Gatenby, who presides over the nation's first mathematical oncology division, located within Tampa's Moffitt Cancer Center, is the last person I would expect to be interested in farming. Pale and bookish, he looks like someone whose lifestyle is mostly internal and sedentary, with long hours spent solving equations on dry boards or staring at backlit boxes in the hospital's radiology viewing room. But as I listened to Gatenby's story, I understood how he came to adopt such an unconventional approach to studying cancer and why, despite his "allergies to everything green," he is so interested in what agriculture has to offer.

"I hated medical school," he explained. "I had twelve years of Catholic school, and it reminded me of catechism. You receive dogma and you recite the facts."

Then, in 1991, having just finished his radiology training, he took a job at the Fox Chase Cancer Center in Pennsylvania. "Once you get involved with a disease, you want to make a difference. The fact that I was not an oncologist, I think, was an advantage. Sitting there in the radiology department, I saw all these tragic cases, and the cut-it-out or nuke-it approach was clearly not working."

Gatenby began to read the scientific literature on cancer and, like Reid, came to the conclusion that the predominant approaches to cancer management had serious shortcomings. So in 2007, when he was offered the job of revamping the radiology department at Moffitt, he accepted on the condition that he could dedicate a portion of his workweek to building and directing a mathematical oncology "collaboratorium," which would take a multidisciplinary approach to understanding and treating cancer.

It was clear to Gatenby that anyone he recruited to his center could not simply adhere to traditional linear scientific reasoning.

"You see, so many processes in the body just do not behave that way," he explained. "In a linear system, if you put in X and get out Y, you assume that if you put in 2X you will get out 2Y. But in a nonlinear system you could get something totally different. You need a much more complicated model for understanding it."

As we talked, I leafed through one of Gatenby's papers published in 2010, "A Theoretical Quantitative Model for Evolution of Cancer Chemotherapy Resistance." It was filled with curving graphs and differential equations—all meant to capture the behavior of cancer cells in a human body. Clearly this man had way more mathematics under his belt than the average doctor. (Later he told me that he was largely self-taught in this discipline.)

"But my biggest insight," said Gatenby, "happened one day when I was surfing the Internet and I came upon the story of the Diamondback moth." This invasive pest was first identified in the 1850s in Illinois, but eventually spread throughout the country and then around the world.

"The moth destroyed cabbage, and since I hate cabbage, my first instinct was to cheer for the moth." Gatenby continued. "Of course, the farmers did not share my opinion. They threw every nasty chemical that human beings could come up with at that insect, but it only became more resistant, evolving to overcome these chemicals. I realized that many of our cancers follow the same story within our bodies. So I started to ask myself: how do farmers these days deal with invasive pests? And I discovered that, in fact, a lot of them are much less linear in their thinking and have many more evolutionarily enlightened therapies than we do in medicine."

Farmers refer to this approach as IPM, or integrated pest management.[1] When farmers use IPM, Gatenby explained, they account for all the dynamics in the system—pests, plants, soil, air, water, weather, and so on—so that crops thrive while pests and weeds are kept at bay.

THE FARMER WHO THINKS LIKE A MOUNTAIN

Winemaker Andrew Mariani is a good example of an enlightened farmer—not that this is immediately apparent. When I first met Andrew, I wasn't sure if he really was a farmer at all or if he just played one in the movies. We were introduced by a mutual friend, Varun Mehra, at a New Year's Day party—certainly not the way I met the other farmers in this book. Varun pointed him out, saying, "That's Andrew. He throws great parties and is a really innovative farmer."

I took one look at this twenty-something with shoulder-length hair, Adriatic good looks, and skinny jeans and felt dubious about the second assertion. (The discovery that he'd been profiled in several glossy magazines did little to strengthen my credulity.) But that night I talked to him long enough to realize that he was not just another hipster trying on farming like the latest wash of denim. Andrew mentioned that he used newer approaches to pest management in his Sonoma vineyard and was drawn to a simpler, old-world form of winemaking with minimal filtration and no added chemicals or commercial yeast preparations. I liked how he described his farm, Scribe, without bragging or embellishment. It also seemed promising that it was a family business and that he came from a long line of farmers. So I decided to pay him a visit.

A week later I was headed north to Scribe, following a two-lane road that winds between dozens of vineyards—wine grapes now being the number-one crop in California's Sonoma County. Not long ago, much of this land was dedicated to other food products, including olives and the locally celebrated sweet and tangy Gravenstein apple. But in the past several decades thousands of acres have been torn up and the fields replanted with the much more profitable pinots and cabernets that now cover the landscape.

The history of Scribe is different. It had been a vineyard before 1919, and then, with the advent of Prohibition, the grapes were uprooted and it became a conventional turkey farm and later (according to rumor) a marijuana farm. When Andrew Mariani and his business partners bought the property in 2007, his mission was to return it to its pre-Prohibition agricultural function.

As soon as I came around the bend and got my first glimpse of Scribe, nestled perfectly in a pocket of hillside, I sensed this difference. Apparently so did a photographer whose work I had seen at a recent wine exhibition at the San Francisco Museum of Modern Art. The display consisted of four photographs, each taken at a different winery. Three of them looked like variations on a vineyard Disneyland, with acres of ruler-straight, perfectly manicured vines, limos in the foreground, and a fortress of a winery. One building had a faux-Pompeian theme complete with columns; another, made out of glass and steel, looked like a mini-replica of the Guggenheim Bilbao. But the fourth photo in the series stood out. There was no monument, only happy-looking couples seated on a knoll, at a wooden table, under oaks, drinking wine. The grass beneath them gave way to slightly irregular rows of wild-looking vines that stretched out in all directions. This photo was taken at Scribe.

Andrew greeted me at the tasting room and led me to the knoll, where he graciously offered me a taste of his 2009 Chardonnay. It was early January, and squinting against the low winter sun, I surveyed his dormant vines. The rows looked unkempt, just as they had in the picture; tufts of weeds were everywhere.

Our conversation was wide-ranging; we talked about food and farming and art and family. He told me that he spent a happy, albeit somewhat uneventful, childhood amid walnut and almond groves in the town of Winters, California, where his family owns Mariani Nut Company. I also learned that Andrew was not the first winemaker in the family. In fact, for generations his Croatian ancestors lived and farmed on the island of Vis in the Adriatic Sea, referred to locally as "Wine Island." In 1904, his great-grandfather Jakov emigrated to the United States, in large part because of a run-in with one particularly destructive aphid called Phylloxera that attacked the grape vines on Vis and most of the rest of Europe. Andrew laughed and pointed out that this is a story of mixed blessings. If not for this microscopic louse, Jakov would never have left beautiful Vis, and two generations later Andrew's father would never have met his mother, who is of Portuguese descent. But he also viewed it as a cautionary tale about what can happen if you grow one crop to the exclusion of all others.[2]

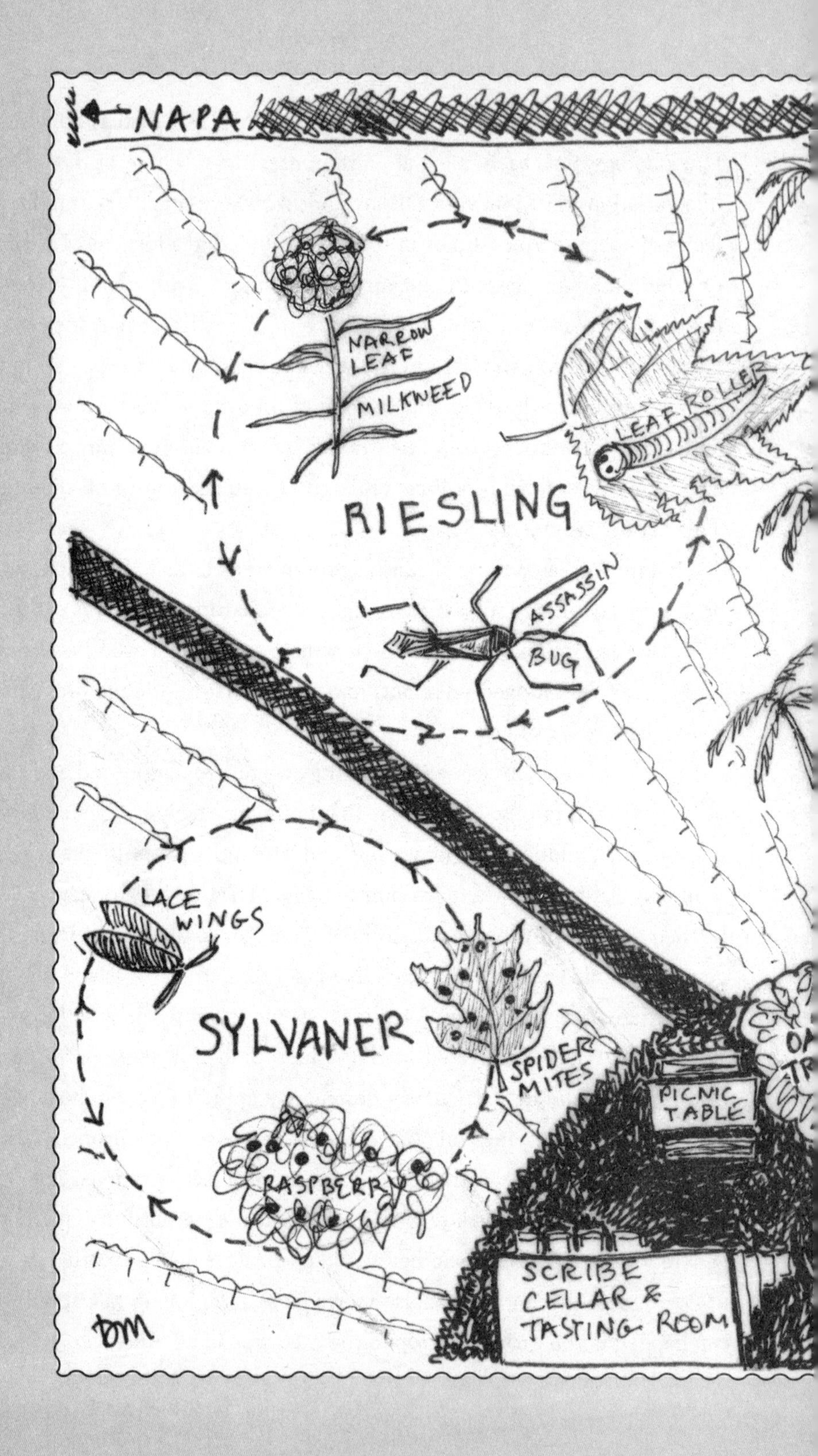
NAPA
NARROW LEAF MILKWEED
LEAF ROLLER
RIESLING
ASSASSIN BUG
LACE WINGS
SYLVANER
SPIDER MITES
PICNIC TABLE
RASPBERRY
SCRIBE CELLAR & TASTING ROOM
DM

Scribe Winery

When I asked him why he chose to farm at Scribe and start making wine, he became serious.

"For years I've been especially drawn to the challenge, the potential of winemaking, because I can't think of a food that offers a more perfect window into what's really happening out there in the field, " he said, pointing to his dormant vines. "Of course, there's certain formulas you can put wine through, but nothing is as beautiful as when you let wine reflect the essence of the place."

Offering me a second glass of the remarkable 2009 Chardonnay, he explained that in a sip of wine, discerning taste buds can experience everything that influenced the winemaking process: the water, the soil, the rocks, the bugs, the plants that grew near the vines, and the yeasts and fungi native to the grape, as well as the chemicals that were used or not used at each phase of the production process. He chose the name Scribe because he viewed wine as a means to tell the story about this land—and his ultimate goal was to revive a natural ecosystem so that his wines would tell a wild and exciting story.

Once again, here was a farmer who chose to design his farm based on a pasture rather than a factory. However, unlike Erick, Cody, and Matt, all of whom had a clear model for accomplishing this, Andrew said that when he first began to cultivate Scribe, he had plenty of energy and time but no specific plan. He spent months clearing away brush and planting vines. Then one summer day, he was dismayed to find that dozens of rattlesnakes had slithered down from the hills above the vineyard in search of water. This discovery, and the events that followed, marked a turning point in the management of the vineyard, one that Andrew refers to as his "Aldo Leopold moment," a reference to Leopold's classic essay "Thinking Like a Mountain."

Leopold writes about how, as a young man, he shot a wolf. "I was young then, and full of trigger-itch; I thought that because fewer wolves meant more deer, that no wolves would mean hunters' paradise." But he quickly realized that the mountain where this wolf roamed saw things quite differently. From the mountain's perspective, killing predatory wolves sets off a chain of untoward events: first an abundance of deer, then a landscape stripped of every edible bush, and finally a mountain littered with "the starved bones of the hoped-for deer herd, dead of its own too much." That was when Aldo Leopold learned to think like a mountain.

Similarly, Andrew could not imagine a good use for a rattlesnake and set about decapitating as many as he could find. But then he noticed something even more unsettling than the occasional rattler underfoot: Tiny burrows had begun to pockmark the newly planted vineyard and his grape seedlings were being attacked from underneath by some invisible, deadly foe. It turned out to be an army of gophers, no longer kept in check by their natural enemy, the rattlesnake. Immediately Andrew stopped killing snakes, and his baby vines flourished. Looking back, he saw this as his first conscious step toward practicing integrated pest management.

TO DO IPM, ONE MUST WALK THE FIELD

On a perfect midsummer day, I headed back to Scribe, accompanied by my son Emet. We had an appointment to meet Jeff Wheeler, the viticulturist hired by Andrew Mariani and his partners to mastermind their IPM strategy. Jeff was sitting under the oaks hunched over a stack of notes, but when we approached he jumped up and offered me a sturdy handshake.

"You might look out and see a vineyard," Jeff said, turning to survey the scene below us, "but what we're really doing here is designing a very complex insectory." He had studied agronomy at UC Davis and had lived in the Sacramento area for years, but his accent still had a touch of his native Oklahoma.

I followed one of Jeff's long fingers as it traced a line from uphill to downvalley, identifying all the components of his insectory: the 150 acres of protected oak; the groupings of native plants; the vegetable garden; a gaggle of free-range chickens and ducks; owl boxes strategically placed high above the fields; the vines themselves; and finally, a jumble of weeds covering the ground between each vine row. Jeff explained that in order to produce delicious wine, he had to understand how all these zones were connected so that he could help beneficial flora and fauna thrive while suppressing those that hindered the process.

Jeff and I headed down to the vines so that he could show me IPM in action. By now, Emet had discovered a rope swing tied to one of the tallest oaks on the knoll, and he soared high above our heads, then out over the vines.

"To do this right," said Jeff, "you first have to monitor what's going on in the vineyard very closely. I could easily manage some of the more conventional operations around here from a rolled-down truck window and a computer, but here at Scribe you really need to get out and walk a heck of a lot more than at other vineyards." He pointed out the weeds underfoot. Now that I was standing between the rows of vines, I recognized a mesclun of greens, reminiscent of what Cody had growing in his paddocks: alfalfa, clover, yarrow, fennel, vetch, buckwheat, and dill.

Jeff Wheeler bent his tall frame forward until his head was at vine level.

"Now, you see, I'm happy because these weeds are sucking up water. Right now, with all the late rains we had, things are so wet that if we didn't have them here the vines would get too lush. That's fine if you're in the business of making dolmas [stuffed grape leaves], but not good for wine."

He referenced his teacher at Davis, Andy Walker, who always said that vines need to be a little dry and stressed in order to make a complex wine.

"Wet vines are also more prone to botrytus [rot]," he continued, "so by pulling away the water, the weeds are actually acting as antifungals."

In addition to their water-diverting properties, the undergrowth served as a lure for "beneficials," insects like ladybugs and wasps that are natural enemies of the pests that attack the vines.

"Still," Jeff chuckled, his hands grazing the tops of the closest weeds, "ten years ago it would have been unbearable the amount of razzing we would have taken for having these growing here. But things have changed."

Jeff delicately turned over a grape leaf and continued his detailed assessment of the state of the vineyard. He pointed at a cluster of tiny reddish pimples on the leaf's underside.

"Okay, so I have two or three spider mites trying to destroy this leaf.... But look there." He redirected my attention to several insects with diaphanous, cornflower-blue wings. This was a bloom of lacewings, attracted to the vineyard by nectar from the weeds and nearby blackberry bushes. "These are beneficials and will devour the mites. So we're okay.

"You see," concluded Jeff, straightening up and striding up the row, "this is why we really need to get out here and look at things up close."

MONITORING CANCER: WHAT HAPPENS WHEN YOU START TO WALK IN THE FIELD?

At a follow-up visit, Dava asked me a very difficult question. "So what about that small percentage of people with Barrett's who go on to develop cancer? Is there a way to predict whether I might be one of them? Or is it all just a big crapshoot?"

My guess is that Brian Reid, the gastroenterologist in Seattle, would find "crapshoot" to be an apt description of the system currently in place for identifying who is truly at risk for esophageal carcinoma. When a patient like Dava undergoes a biopsy of her esophageal lining, that biopsy is sent to a pathologist whose job is to look at the tissue under the microscope and decide whether the cells in question show any "dysplasia"—changes characteristic of cancer. But Reid told me that, based on a series of studies, he has come to view this assessment as surprisingly subjective or "operator-dependent." The way he described it, the pathologist's analysis seemed more art than science—like an antiquarian using motifs and shapes to date the provenance and era of an ancient vase.

"In general," he said, "the diagnosis of dysplasia, or even high-grade [advanced] dysplasia, often does not correlate with esophageal adenocarcinoma." He added that the widely accepted practice of looking for specific genetic markers for malignancy can also be an unreliable way of identifying who will ultimately get cancer. Fortunately, he and his colleagues were working on developing much better markers for identifying those who are truly at risk for developing the disease.

Shortly after my second visit to Scribe, I had tea with Carlo Maley, who is helping Reid identify these predictive markers. Maley, whom I met in a café near his offices at the UCSF Mount Zion Cancer Research Center, is another member of Reid and Gatenby's evolutionary oncology circle. An evolutionary biologist by training, he first became interested in the connections between evolution and cancer in the late 1990s when he was a postdoctoral fellow. At that time, he entered the words "evolution" and "cancer" into PubMed, the premier bioscience search engine, and came up with a measly six hits. Soon after, he was introduced to Brian Reid at an interdisciplinary workshop in Santa Fe, and the two men began to collaborate.

At this point Maley had examined hundreds of the esophageal biopsies from Reid's patients, collected both from the time they were diagnosed with Barrett's and from their periodic follow-up endoscopies. Like Reid, he viewed Barrett's as an ideal model for understanding the natural progression of precancerous cells, precisely because the outcomes are not muddied by the drugs and surgeries that are lobbed at most other early malignancies.

Maley told me that he and Reid, either separately or in collaboration, had identified a new set of measures to assess a tumor's metastatic potential, measures that he felt are more individualized and offer a more dynamic representation of tumor behavior than what is currently used by oncologists. Their novel approach includes assessing the percentage of cells with an odd or abnormal number of chromosomes—also referred to as "degree of aneuploidy"—the speed of cell reproduction, and the total number of different mutations within one tumor. Reid's lab had also discovered that individuals who are more at risk for cancer have biopsies that show loss of the p53 gene. This gene is often referred to as the guardian of the genome, since it acts as a watchdog, detecting DNA damage and killing mutant cells, and when it disappears from the biopsies of someone with Barrett's, it may indicate a fifteen times greater likelihood of developing cancer. As I sipped my tea and listened to Maley describe the wide variety of markers they were using to better understand a tumor's behavior, I was reminded of how Jeff Wheeler also relied on a host of factors to assess the state of the vineyard.

Maley hoped that eventually their work will help eliminate what he called "the overdiagnosis, underdiagnosis syndrome": thousands of patients with benign-type Barrett's being subjected to unnecessary worry, expense, and potentially harmful procedures while a select few with more aggressive cells are not alerted to the fact that they are especially at risk. He wanted a system in which it's possible to identify and focus prevention efforts on those who really need it.

I passed on my newfound knowledge to Dava, who listened intently. But even as I spoke, I could anticipate her next question, and I was not looking forward to answering it. She had her records in front of her, and none of the tests I mentioned had been performed on her esophageal tissue.

"So who can do these tests on me?" she asked.

With the exception of flow cytometry to measure the number of aneuploid cells (which is now being done in Reid's clinic as well as in a handful of other Barrett's specialty centers around the country), all of these tests are still experimental. They need to be studied in more patients with Barrett's before they can become a part of standard clinical care.

Given that Reid, Gatenby, and Maley's work spans several decades and has been published in many reputable journals, one has to ask why it's not being replicated and expanded upon in cancer research centers nationwide. As I understand it, one major obstacle is a general resistance to getting out of that metaphorical truck and walking the field. Jeff Wheeler at Scribe understands that no two fields are the same and that he needs to stoop between the rows, turn over leaves, feel the soil, and assess the condition of his neighbor's fields in order to decide whether his field is at risk. Similarly, no two precancer or cancer patients have cell populations that behave exactly alike. For oncologists to truly assess the risk of each individual, they need to stop relying on a few static markers and make sense of a variety of data, much of which is dynamic. But as Reid told me, "Many [medical colleagues] are frightened by that. I'm more frightened by the current situation where we continue to do things that don't work."

THE EVO GAME MASTER

That day at Scribe, Jeff Wheeler seemed quite pleased with the vineyard's ratio of pests to beneficials, but I asked him whether he would have resorted to synthetic pesticides or herbicides if he had found an unacceptably high number of mites or other pests.

"Well, now," he said, "if I were to go straight to the synthetic herbicides and pesticides to deal with an imbalance, I could run into all kinds of problems. If you don't believe me, just look at those FRAC [fungicide resistance action committee] code charts.[3] They have to update them all the time because there are always new resistances. You see, most chemicals attack one specific protein in the pest, so with one easy mutation that pest can evade its attacker." He saw the same potential for resistance in the genetically modified or GMO root stock varieties that are currently being developed and tested in viticultural research centers around the world. Most GMO plants have a gene that blocks a discrete enzyme or one step of viral or fungal metabolism; based on everything he knew, Jeff was certain that it won't take long for the offending pest or disease to outsmart the transgenic technology.

It's important to understand that Jeff came from the heartland of Big Ag and that he did his graduate work at a university that still receives a substantial amount of funding from Monsanto, Dow Chemical, and other corporations invested in industrial farming. He also had a lot of vineyard clients who, unlike Andrew, encouraged him to use whatever means necessary to make pests and weeds disappear. So when he said something anti-chemical or anti-technology, one had to assume that his comment was highly considered and not just the knee-jerk response of a weed-hugging liberal.

Jeff explained that a better way to combat pests and weeds is to use weapons that cannot be easily circumvented with a simple mutation. That way, in the rare instances when the pest evolves sufficiently to evade all the obstacles set forth by its attacker, the effort is so costly to its biological system that one of two things will happen: It will no longer effectively reproduce, or it will become especially vulnerable to another form of pest control. He referred to this as creating an "evolutionary double bind." Jeff viewed integrated pest management, at its best, as the creation of a series of double binds, and he gave me specific examples of how he implemented this approach at Scribe.

First, he grew the plants that would attract beneficials (such as mite-fighting insects) into the vineyard.

"Think about it," he said. "It's pretty hard for a mite to mutate so much that it can overcome the attack of a wasp or a lacewing." That made sense to me. A mite's susceptibility to wasps is not encoded on one protein or one short length of DNA; it's something much more nonspecific. Really, the only way a mite could evolve enough to evade an attacker moth would be to lose its attraction to grape leaves and never enter the vineyard in the first place. Chickens pecking their way through the vineyard, snakes in the grass, and owls roosting in boxes overhead are other examples of enemies that a pest is unlikely to evade with one or a few discrete mutations.

In addition to creating an environment that was welcoming to beneficials, Jeff tried to foster conditions that were inhospitable to the pests themselves. Mites, for example, thrive on dusty leaves, so it's important to control loose, dry topsoil in the vineyard. To do this, he advised the vineyard crew to routinely mist the grape leaves with water and support the growth of dust-dampening weeds between the rows. (Yet another role for those weeds!)

As we talked, I watched Jeff pull some leaves off of one of the lower vines.

"What I'm doing here is called leaf-thinning. This is another kind of environmental manipulation, since it gives the plant and the fruit room to breathe and offers a chemical-free way to prevent mold."

When I had met with Andrew, he mentioned intercropping with fava and other legumes as one more way to safely control pests. His great-grandfather's experience with the devastating Phylloxera made him all too aware of the pitfalls of monoculture farming and he was determined that his fields maintain biodiversity so as to not serve as a beacon, attracting overwhelming numbers of one kind of pest. Intercropping with "nitrogen-fixing" plants such as fava also offers an efficient, nonsynthetic way to replenish nitrogen in the soil.

If a pest population began to expand, Jeff would set pheromone traps: small, tent-shaped bags filled with female bug scents. Suspended from vine stakes, they act like Sirens for male bugs, luring them in as they search for a mate and then making it impossible for them to escape. Given that pheromone traps are essentially sexual trickery, any mutation that allowed a pest to ignore the trap's beguiling scent would most likely interfere with its drive to mate and reproduce. Definitely an evolutionary double bind.

Exotic beneficials or non-native insects that prey on a certain pest are yet another part of an integrated pest management tool kit. They can be purchased from a vineyard supplier and released in a problem area within the vineyard. Jeff kept this higher, more intense level of field management in reserve, as one can never fully anticipate the consequences of introducing a foreign beneficial into a field. Far too often, imports intended to maintain the natural ecology of a farm have turned out to be pests themselves.

As a last resort, Jeff turned to the judicious and targeted use of organic pesticides. Sulfur was one of his favorites because it injures many sites within arthropod cells and therefore acts more like a wasp than a chemical. Sulfur is also a naturally occurring compound, and its residue disappears quickly from a field, whereas synthetic chemicals like Glyphosate (commercially known as Round-Up) remain there for weeks, destroying beneficials and having far-reaching toxic effects within the ecosystem. Jeff explained that, to a lesser extent, sulfur can also be toxic to beneficials, so when he resorted to this compound he was careful to apply it at times when it would do a minimum of damage. For a heavy pest burden, he also sprayed natural oils and soaps on leaves. These prevent bugs from attaching without negatively affecting the health of the surrounding environment.

As I listened to this hierarchy of interventions, I was impressed with how logical it all sounded. Jeff Wheeler's therapies started with the most gentle and he escalated his interventions based on the specific needs of each field. His initial strategy was to promote beneficials, intercrop, and control moisture and dust. Then, if necessary, he used pheromone traps and exotic pests. Finally, if things were still not right, he would turn to bigger guns such as sulfur and oil, but only in small doses and only when truly needed. Clearly, in the case of Scribe, this considered approach had helped Andrew Mariani achieve his ultimate goal: keeping the vineyard healthy and making a wine that reflected this vitality.

IPM IN CANCER CARE

Bob Gatenby's ultimate goal, on the other hand, is to keep people healthy and to prevent them from dying of cancer. But the four new strategies he proposes for achieving this are strikingly similar to those used at Scribe.

Strategy #1: Boosting Beneficials

Borrowing from IPM principles, where the priority is to optimize conditions for pest-fighting beneficials, Gatenby wanted to stimulate the immune system, allowing it to fight cancer cells. He listed sleep, exercise, a vegetable-based diet, and stress management as low-tech examples of immune boosters, while more sophisticated interventions include immunotherapy and oncolytic (cancer-killing) viruses. Gatenby acknowledged that so far studies using these two agents have done little to stem tumor growth, but he thought that would change once researchers found a way to promote immune cells that could specifically prevent cancers from accessing a nutrient source.

"Hawks are particularly effective when they exert the most threat in habitats where mice must seek food," Gatenby explained, once again referencing the natural world in order to illustrate a difficult biomedical concept.

Gatenby and Maley are also interested in the potential for benign cell boosters. Similar to the blackberry bushes and plum trees that border the vineyard at Scribe, planted specifically to attract beneficial insects so that they can outnumber the pests, benign cell boosters are substances that promote the growth of normal cells and allow them to outcompete the typically faster-dividing cancer cells. One study now under way in Maley's lab has offered some early evidence that vitamin C, at least in a petri dish, may act as such a booster. Other boosters are likely to be identified as cancer researchers pursue this novel path of inquiry.

Strategy #2: Pheromone Traps and Sucker Gambits

Gatenby also believed that creating evolutionary double binds for cancer cells is an excellent way to selectively control their growth and replication. He and his colleagues were exploring how to make a cancer cell mutate so that it can no longer reproduce or so that it becomes more vulnerable to a chemotherapeutic agent. He told me about a somewhat accidental double bind uncovered while studying an experimental vaccine to promote p53 production in people with small cell lung cancer. (Remember those p53 guardian cells mentioned earlier?) Although 57 percent of the people who received the vaccine saw an increase in p53 production, their tumors continued to grow unabated, so the researchers deemed the experiment a failure and they switched the patients to a standard chemotherapy regimen. But then things got interesting. On this second round, 62 percent of the prevaccinated patients went into remission, a much greater number than the 8 percent remission rate typically seen with this chemotherapy regimen! Gatenby explained that the initial vaccine somehow primed the tumor cells so that they were more vulnerable to the follow-up treatment. (Sadly, by the time researchers realized that there might have been a synergy between the vaccine and the chemotherapy, the vaccine manufacturer had gone out of business and the innoculum was unavailable to other patients. Gatenby lamented that this scenario occurs far too often with promising therapies.)

Strategy #3: Leaf-Stripping and Environmental Control

Gatenby and his evo cancer colleagues often refer to the "seed and soil hypothesis"—the idea that a malignant cell needs a supportive milieu in order to flourish, just as a weed can only take hold in the right conditions. Many of their studies focus on uncovering environmental factors that control cancer formation and growth. They already know that cancer cells are stimulated by sugary, low-oxygen (acidic) environments with poor blood flow—all conditions where normal cells do poorly.

When I asked Gatenby how this information could be put into practice for someone with cancer or at high risk for cancer, he recommended a low-sugar, alkaline diet with lots of vegetables and a minimum of meat. (He was currently studying whether alkalinizing agents such as antacids and dairy can also be beneficial.) He mentioned that remaining fit and lean and avoiding tobacco also discourages cancer growth. Finally, he referred to a paper published in *Cell*, one that I first read when I was learning about stress in chickens. In this study, mice were injected with either melanoma or colon cancer cells and then randomly placed in either furnished cages (with lots of activities) or empty ones. In general, the mice in the furnished cages had slower-growing tumors than their empty-cage counterparts, and the tumors in some mice with furnished cages disappeared altogether. Of course, this study has not been replicated in humans, but Gatenby suspects that immersion in a mentally and physically stimulating environment—as opposed to a life spent on the couch—might help prevent, contain, or reverse a cancer.

Strategy #4: Tailoring the Treatment to the Field

Reiterating what Brian Reid had already told me, Gatenby explained that standard cancer therapy is based on a simplistic, one-size-fits-all approach: diagnose the cancer, decide on the chemotherapy (or radiation) regimen, then give the patient the maximal tolerable dose (MTD). Repeat this treatment X number of times at Y intervals, then perform an imaging study to reassess.

"But here's the problem," said Bob Gatenby. "This standard approach primarily targets the chemically sensitive tumor cells while sparing those that are chemo-resistant." In the same way that field pests listed on the FRAC chart quickly overcome their pesticide foes, these resistant cells grow back in force, invading other tissues as metastases and eventually killing their host. (This explains why all too often my patients are declared "cured," only to succumb to a cancer recurrence five to ten years down the road.) To avoid this, Gatenby proposes a radically new approach called "adaptive therapy." This system is designed to treat cancer cells just enough to keep them at bay and yet not so much that the treatment selects for the uber-resistant cells. Like a good IPM practitioner, Gatenby believes that constantly surveying the conditions in the field and maintaining a delicate balance is the key to preventing a full-blown infestation.

So far adaptive therapy has been used only to study tumor behavior in mice, but here's how it works: Administer an initial dose of a chemotherapy that's slightly lower than the standard dose and then reassess in three days. If the tumor has not changed in size, give no more of the drug. If it has increased by 10 percent or more, then give the same dose again. But if it continues to increase at the next measurement, then increase the dose by 10 percent. If, on the other hand, the tumor is stable for two consecutive measurements or shrinks in size, then decrease the next dose by 10 percent. Continue to do this every three days. What Gatenby found in his mice (and what his mathematical models predict for humans) was that this tailored approach can keep a tumor static indefinitely while minimizing the use of potentially harmful chemicals. By contrast, rodent tumors treated with the standard MTD method increased in size and eventually metastasized.

In 2010, Gatenby was given a grant by the McDonnell Foundation to start testing adaptive therapy in humans, and recently he and Carlo Maley have jointly received even more funding. Impressively, this latest award was from the National Cancer Institute, under its "Provocative Questions" initiative.

A NEW TAKE ON CANCER

While Bob Gatenby and I sat in the cafeteria at Moffitt Cancer Center, eating lunch and discussing cabbage moths, evolutionary game theory, and adaptive therapy, it occurred to me that what he and his colleagues were proposing was not just a new strategy for cancer care and prevention. It was an entirely new way to think about the disease. In general, cancer is perceived as a frightening invader that must be eradicated before it kills its host. Gatenby, on the other hand, saw it more like a pest in an integratively managed field—it will always be there to some degree, but not so much that it overwhelms the beneficials and destroys the crops. In this view, in other words, cancer is a chronic challenge that must be contained and only sometimes reversed. And as we have seen with adaptive therapy, striving for containment rather than eradication is more likely to control the disease in the long run.

This is not how my patients hear cancer portrayed in most oncologists' offices or in their support groups, both places where words like "fight" and "conquer" are commonplace. Nor is it how cancer is discussed in obituaries, where people who die from the disease are described as having "lost their battle." Seeing cancer more as a chronic pest and less as an invader may be an important first step to better understanding the disease. Perhaps only then will our approach to cancer management be as enlightened as what Jeff Wheeler is practicing in his fields.

WOLVES, SNAKES, AND BARRETT'S

Peering ever more closely at Barrett's esophageal cells through their high-powered electron microscopes, Brian Reid and his colleagues noticed something remarkable: Those transformed cells, the ones considered unstable and precancerous, actually look pretty darn normal given the role that they are expected to play in the lower esophagus. In fact, the more they studied these cells, the more they concluded that they were perfect mucus- and bicarbonate-producing factories—exactly the kind of cell you want to have lining an organ that's constantly bathed in acid. By using a technique called flow cytometry, they also discovered that in more than 95 percent of patients with Barrett's the chromosomes in these cells are diploid, meaning they exist in pairs just like the normal, somatic cells of the body. From these observations, Brian Reid came to suspect that, far from being the enemy, the majority of Barrett's represents a normal, healthy response to having acid reflux. And since Barrett's tends to run in families, he saw this as a fascinating example of evolution playing out on the level of the organism and the cell. In other words, families predisposed to reflux pass along a genetic trait to make Barrett's-type cells, and cells exposed to acid evolve to work like a mucus-producing part of the intestine.

Listening to Reid, I began to wonder if the protective potential of certain Barrett's cells represents a larger phenomenon. Are there special guardian cells within other types of cancers and precancers that have yet to be identified? If so, indiscriminately killing anything that is "not normal" might lead to unintended outcomes, just as trying to eradicate the wolves or the rattlesnakes produced untoward results.

When I explained all this to Dava, she looked annoyed. This was not at all what she was expecting. She had asked about interventions that would blast her precancer, and here I was telling her about the protective role these cells might be playing in her esophagus. But after a considerable silence I saw her expression soften, and she actually leaned back in her chair. It was at that moment, she told me later, that she had her first inkling that this was not a war but something much more nuanced. Perhaps those Barrett's cells, like her liver and heart, were actually a vital part of her.

A month or so later, I saw Dava again. Based on the idea that more antioxidants and a higher pH environment were less conducive to cancer growth, she had upped the polyphenols in her diet by eating more vegetables, fruits, and fresh spices (such as, thyme and oregano) and was avoiding alcohol and red meat. She had also started taking a non-steroidal anti-inflammatory drug (NSAID) or two each week, based on the discovery by epidemiologists in Brian Reid's lab that Barrett's patients who took ibuprofen or aspirin at least once a week developed esophageal cancer at one-third the rate of those who never took NSAIDs. (People who took NSAIDs for a while and then stopped had a risk somewhere in between these two groups.)[4]

Dava also began having dinners that were on the light side, with soups, grilled vegetables, grains, and very little meat or dairy, and she changed her schedule so that her last meal was at least three hours before bedtime. On my advice, before each meal she drank a glass of water with a tablespoon or two of apple cider vinegar or lemon. Although these are both acidic, they are also potassium-rich, and potassium neutralizes acid in the esophagus. She wasn't sure which of these changes made the difference, but her heartburn finally improved.

At some point, Dava plans to consult with Dr. Reid in Seattle. Meanwhile, getting more accurate information about Barrett's has been so reassuring that she's hardly thought about the issue since our last appointment. In fact, she feels calmer than she has felt in years. I wondered if this too could explain the fact that her reflux symptoms had practically disappeared.

HARVEST

It's 5:45 on a warm mid-September morning, and I'm driving a bit too fast along that curvy one-lane road to Scribe. Andrew's text came the night before: "Harvest Sylvaner tomorrow 6 AM." For days Andrew, his younger brother Adam, Jeff Wheeler, and other members of the Scribe team had been sampling the grapes, preoccupied with data about brix (sugar) level and pH. Finally, the time had come to pick their crop of Sylvaner, an ancient variety of white grape that's rarely harvested in the United States but was grown in the original vineyards at Scribe and also happens to be popular in Andrew's ancestral Croatia. I was rushing because I instinctively felt that an invitation to such an important event deserved complete punctuality.

It was still pitch-dark when I arrived at the oak knoll. I got out of my car and listened. Silence. Then, in the distance, down in the vineyard, I heard muffled voices and the low hum of an engine. I headed down the hill toward these sounds and eventually spotted a few pinpoints of light dancing among the vines. Getting closer, I was amazed to see at least a dozen people silently working within a thirty-foot segment of vine row. Out front were Scribe's regular employees, the city-raised, aspiring farmers I'd previously seen tending the gardens and the tasting room. This morning they were the unskilled leaf-strippers who gently pulled back leaves to reveal the perfect bunches of ripe grapes. Following them were half a dozen contract laborers, mostly Mexican, who used their curved blades with a surgical finesse to separate the fragile grapes from the enveloping vines. They glided down the row in a crouch, filling bin after bin with the golden Sylvaner. Presiding over this were the Mariani brothers. They stood on the tractor hitch and used their headlamps to carefully examine each bucketful of fruit as it was dumped into a half-ton collection wagon.

As I watched them, I remembered Andrew telling me that in every wonderful bottle of wine there are at least a hundred little decisions, and I realized that what I was seeing was not just quality control but one such decision. By selectively pulling out leaves and stems from the collection wagon and deciding how many sweet fungal spores (botrytus) to leave behind, Andrew and Adam were literally deciding how much of the terroir, the environmental influences, should end up in the fermentation tank. This was extra important with Sylvaner because it's a mild-tasting grape, not one destined, in Andrew's words, to become a "big-ass 100-pointer"—a reference to Robert Parker's highly influential rating system. Because of its mellowness, it would reveal all the flavors of Scribe: the weeds, the beneficials, the neighboring plantings, the botrytus, the sun exposure, the water, the rattlesnakes, and all the many choices Andrew Mariani and Jeff Wheeler had made throughout the season.

By 9:00 A.M., I had spent a meditative three hours stripping leaves, and it felt satisfying to know that three tons of Sylvaner were on their way to the nearby winery to be pressed. I climbed into a pickup with Andrew and his uncle (a co-founder of Scribe also named Andrew), and we followed our harvest to its destination. We stood around for a while chatting and playing with our cell phones while workers at the winery checked lot numbers, filled out health department forms, and prepared the presses. It seems that standing and waiting is something you get good at if you make wine. My 5:00 A.M. wake-up time was starting to catch up with me and I desperately wanted a cup of coffee, but Andrew warned me that it would mess up my taste buds, so I abstained. Finally the forklift hoisted the first half-ton of grapes high in the air and dumped them into the mouth of a stainless steel press. There was a loud squashing sound, and I could hear grape juice cascading into the catchment bin. The moment that we had all been waiting for had arrived.

Andrew offered me a wineglass of the thick, tawny juice and I took a sip.

Then I stood there and considered what I'd just tasted, recalling a letter Andrew recently sent to members of Scribe's Viticultural Society, or wine club, myself among them:

We are on the cusp of harvest here on the Scribe Estate; by the time most of you read this it will be going full-force. With this vintage, our Estate vines enter their fourth leaf, their largest harvest to date, and this offers us our first real look at the vineyard's long-term potential. There's lots on our minds these days: brix levels, pH levels, barrel programs—or lack thereof—fermentation approaches, etc. . . . but behind all the analysis, planning, action, or inaction is a deeper goal: that with this vintage, like each one before it, we continue to remove the veil of this place, and more fully recognize its Terroir. . . . Our interest in Terroir, and in winemaking, is to build an enduring connection with the landscape, to survive on it, and to live well.

I closed my eyes and took another sip. I thought I could taste a faint bit of anise and dill from the weeds, some iron from the rich soil—and, wait, could that peppery taste be the lacewings? Scribe wine is not cheap (as Jeff told me, "If I'm producing a $10 bottle of wine, I can't afford to come in here and drop clusters"), so committing to buy four bottles every quarter felt like an indulgence. But right then I made a decision: Nine months from now, when I received this vintage of Sylvaner and loaded the elegant green bottles into the racks that house my nascent wine collection, I would put one bottle aside. It might seem a little inappropriate for a doctor to offer wine to a patient—especially a patient with gastroesophageal reflux—but the poetry was too perfect, and I knew Dava would appreciate it.

Community Medicine, One Plot at a Time

Not so long ago public health was the science of controlling infectious diseases by identifying the "cause" (an alien organism) and taking steps to remove or contain it. Today's epidemics have fuzzier boundaries . . . : They are the result of the interplay of genetic predisposition, environmental context, and lifestyle choices.

—Paul Plsek and Trisha Greenhalgh, "The Challenge of Complexity in Health Care" (2001)

STANDING IN the subway station at 130th Street and Grand Concourse in the Bronx, I asked the agent behind the Plexiglas which of the two exits would get me closer to the farm.

"What farm?" she boomed at me through the microphone. "There's no farm up there."

Uncertain, I checked my iPhone. According to Karen Washington's most recent e-mail, this was exactly where I was supposed to be, so I ignored the agent, randomly chose an exit, and headed aboveground. Sure enough, right on the other side of 130th Street, behind a wrought-iron fence, I spotted a sunny expanse of green, complete with sheds and a scarecrow. Vindicated, I wanted to run back to the subway and pry that grumpy, myopic subway worker from her bulletproof booth so that she could see La Finca del Sur, the quarter-acre of food production that she probably passed on her way to work every day.

Instead, I crossed the busy street and peered through the gate. Among the raised beds I spotted a woman leaning on a hoe, a tangle of golden dreads fanning out on her shoulders like a ceremonial headdress. She was deep in conversation with a young man who I would soon learn was Omar, a student at New York University turned farming intern. I called out, getting their attention. "Daphne? You made it, baby!" she shouted back, swiveling to face me. "Just let yourself in."

Unwinding a thick chain, I opened the gate and entered the garden. Almost like magic, the honks and screeches from Grand Concourse fell away, replaced by the rustle of leaves and the tinkle of wind chimes on the scarecrow. It was the fifth of November, and New York had recently been pelted by an early snowstorm, but the day was unseasonably warm and there were still plenty of edibles growing in the raised beds. A cat jogged up and gave my ankles a welcoming rub. Just then, a deafening roar broke the stillness as a northbound commuter train flashed by overhead. The cat didn't even flinch.

FOOD DESERTS AND FOOD MIRAGES

My plan to visit Karen began about six months earlier, while sitting in the conference room of a spanking-new Giant food store on the outskirts of Philadelphia. The store was hosting an annual meeting of retail dietitians, and

I'd been invited to speak about traditional diets. Sitting in the room were representatives of many of the largest supermarkets on the planet, including Ahold, the conglomerate that owns Giant Foods and Stop & Shop; SuperValu, corporate parent of Albertson's and Shaw's; Fresh & Easy; and Walmart. The attendees, all women, were responsible for advising merchandisers about foods that might offer nutritional benefit to shoppers yet would still guarantee a profit. Given that the profit margin on a food increases in direct proportion to how processed it is, their task is a challenging one. Therefore, each year these dietitians convene and share valuable tips on how to sell fruits, vegetables, and other healthy foods within an unhealthy system.

At this meeting, attendees were buzzing about "food deserts." The USDA defines a food desert as a low-income census tract in which at least 33 percent of the population (or five hundred people) live a mile or more from a supermarket that does at least $2 million in annual sales. (Perhaps a better definition is the one given by environmental writer Mark Dowie: "When you have to go twice as far from your house to get a piece of lettuce than a bag of chips.") A map highlighting our nation's food deserts looks almost identical to one showing zones with the lowest life expectancy and the highest rates of obesity, diabetes, and heart disease.

The general consensus among the dietitians was that irrigating these deserts with supermarkets would boost fruit and vegetable intake and offer an excellent antidote to these worrisome disease trends. They were delighted that First Lady Michelle Obama had emerged as an important ally in this effort. Just about a month earlier, she'd launched her "Let's Move" campaign to reduce childhood obesity by presiding over a ribbon-cutting ceremony for a Fresh Grocer supermarket in Philadelphia's inner city.[1]

"Our goal is ambitious," announced Ms. Obama. "It is to eliminate food deserts in America completely in seven years. Tackling the issue of accessibility and affordability is key to achieving the overall goal of solving childhood obesity in this generation. And we saw this example today on our visit to the Fresh Grocer at Progress Plaza."

It was easy to follow the logic inspiring the organizers of the White House Food Campaign, as well as the motivations of the dietitians gathered in Philadelphia. But sitting inside that Giant supermarket, I wasn't quite buying it. For sure, the place was big and new. And it did have a picturesque faux farmers' market, complete with "sunbrellas," conveniently located inside the entry, as well as an education room to host cooking demonstrations. It even offered customers an opportunity to consult individually with an in-store dietitian. But at the same time, I could see why the new market might not be a catalyst for dietary change.

First of all, it was located outside a neighborhood in a strip mall, a place accessible only to someone with a car and enough free time to drive there. Second, although this store did have plenty of fresh fruits and vegetables, nicely displayed at the entry, it still offered aisle upon aisle of the high-calorie, low-nutrient foods that enable retail grocery giants to turn a profit. In order to pick up staples like milk and eggs at the back of the store, shoppers must push their carts past all that tantalizing junk food. In the end, the soda, chips, cookies, and canned food provide a heavy counterweight to the virtuous (and now squashed) bunch of kale that was selected on the way in. Demoralized, I realized that although the retail dietitians at my meeting had the best of intentions, their employers wanted to improve the American diet and fight obesity about as much as a body shop wants to eliminate car crashes.

I was also troubled by the suggestion in recent high-quality studies that a new grocery store in a food desert does little to solve a community's health and nutrition problems.

One such study followed five thousand young adults living in four major cities (Birmingham, Minneapolis, Chicago, and Oakland) for fifteen years and tried to understand how proximity to markets and fast food affected their diet. Controlling for factors such as age, ethnicity, marital status, and income level, the results showed that a supermarket within a kilometer of the home does not improve diet quality nor boost vegetable intake. Other studies from the United States and abroad have reproduced these findings: One year after plugging a grocery gap in the Glasgow area of the United Kingdom, community members reported no change in vegetable intake or in self-rated health.

In short, while it's enticingly simple to think that more supermarkets could offer a panacea for obesity and for our national vegetable deficit, the evidence does little to support this idea.

URBAN FARMING: A PUBLIC HEALTH INTERVENTION?

That afternoon in the Giant company conference room, I tuned out a talk on promoting active lifestyles being given by a dietitian from PepsiCo and began to search PubMed, the online medical database. I was looking for some promising community-based approaches to boosting vegetable intake and combating obesity and other chronic health problems. It was during this cyber-wandering that I first came across a series of papers published by Jill Litt, a public health researcher in the Department of Environmental Health at the Colorado School of Public Health. Litt was studying a food source far beyond the purview of supermarket chains: home and communal gardens. Her research, which included Denver's lowest-income areas, looked at outcomes similar to those measured in the food desert studies. She found that gardening accomplishes what grocery chains cannot: getting people to eat more fruits and vegetables. For example, in her data set, 56 percent of gardeners ate the recommended five or more servings of vegetables per day, while only 25 percent of nongardeners hit this target. These data were impressive!

Litt's articles inspired me to enter the search terms: "community gardening" and "public health" into the PubMed portal, and to my surprise my query produced eighty-one articles. I began to scroll through them. Some looked at the impact of urban gardening on specific health issues, such as diabetes, arthritis, dementia, or depression. While others looked at gardening's effect on behaviors, including alcohol consumption, vegetable intake, or hours of exercise per week. Still others studied more qualitative outcomes, like "self-rated health." Almost without exception each study showed that gardening was linked to improved health.

And while many of these studies had their shortcomings (they were observational as opposed to experimental, so perhaps there was some bias in terms of who chose to participate), many of the results suggested that it's truly gardening—above and beyond other factors, such as educational levels—that makes the difference. In other words, the physical act of growing food leads to all kinds of health benefits, including higher fruit and vegetable intake, lower blood sugar and body mass, fewer depressive symptoms, and increased exercise capacity.

I was excited about what I was reading, but there were a couple of details in these reports that I found perplexing. First of all, many of these studies took place in northern and eastern states, where even the sunniest and most sheltered outdoor garden is likely to be dormant and nonproductive for six or more months of the year. Second, in some of the gardens that were studied, the gardeners were growing flowers and ornamentals instead of food. How could seasonal (and not necessarily edible) gardening generate healthier eating patterns than a supermarket that offered fruits and vegetables seven days a week, twelve months a year? And how could it produce the other positive health outcomes, such as higher "self-rated health," even in times of fallow?

Gazing up at the acoustic-tiled ceiling of the Giant food store conference room, I pictured hundreds of square garden plots, bursting with vegetables. At that moment, I decided to find out why growing plants in the inner city offers such effective preventive medicine. I needed to spend time with an urban farmer.

THE MATRIARCH OF URBAN FARMING

"You must be hungry, baby," Karen said seconds after giving me a welcoming hug. She pulled a small cooler from her bag, led me over to a nearby picnic table, and set out a box of crackers and some salsa that she'd made earlier that morning from her garden's bounty. "Maybe for you this is breakfast?" she chuckled, referring either to the fact that it was still breakfast time in California or that Californians love salsa any time of day. As I scooped up cracker-fuls of tasty late-season tomatoes and peppers (it was a perfect breakfast), Karen told me about how she became a farmer.

Karen Washington's Garden of Happiness Viva Salsa

"Here's a salsa recipe that takes great advantage of local produce at the height of the season. I also love to add a chopped peach or mango to this refreshing, mouthwatering dish." —KAREN

INGREDIENTS:

- ½ cup cilantro, chopped
- 1 tablespoon coriander, ground
- ½ red onion, finely chopped, or a small bunch of scallions (bulbs and greens), chopped
- ½ cup red bell pepper, finely chopped
- ½ cup green bell pepper, chopped
- 1 jalapeno pepper (with seeds for more spice), minced
- 3–4 firm ripe tomatoes
- 1 tomatillo, chopped
- sea salt, to taste
- 1 lemon

INSTRUCTIONS:

Turn on your favorite salsa jam and get ready to mix it up. In a mixing bowl, combine all of the herbs and veggies. Add salt and a squeeze of lemon juice to taste. Stir, cover, and refrigerate for about thirty minutes. Serve with your favorite tortilla chips.

In 1985, days after she moved north from Harlem to what she called "her small piece of the American dream," Karen realized that she'd made a terrible mistake. She never should have invested in real estate across the street from a vacant lot. Like dozens of other abandoned blocks scattered throughout the Bronx in that era, the one that faced her front door was a nest for junkies and rats. Karen watched in dismay as passersby casually tossed stained mattresses and bags of trash over the fence.

"When you're living where people are throwing garbage, people look at you as garbage. Cops would come and, under their breath, call us 'monkeys' for living in garbage."

Then one day, about three years later, everything changed. Karen looked out her window and spotted a man standing in the middle of the lot holding a shovel. She ran outside to ask him what he was doing, and he told her that he was working with the Bronx Green-Up program, an initiative funded by the New York Botanical Garden to turn vacant lots into community gardens. As luck would have it, they had identified Karen's block as one of their very first sites.

"I said, 'Can I help?' " she told me, recalling that day. "And the rest is history."

As we talked, I noticed Karen's sweatshirt. It was standard-issue grandma wear, the kind you buy in an airport gift shop, its front emblazoned with a cartoon drawing of an elderly mama bear wearing a bonnet and spectacles, repairing a teddy bear. Under the image it said in cursive: A GRANDMOTHER CAN ALWAYS PATCH THINGS TOGETHER.

I smiled. Cutesy as it might have been, this was the perfect motto for a woman who had made it her life's work to patch her community together. Over the past two decades, she'd helped transform trash-filled lots throughout the Bronx into verdant refuges. Some of the gardens considered her a founder, others a mentor, and still others an inspiration. One thing was certain: You could see Karen's handiwork everywhere. If you were to visit just one of these sites, you might think it a stretch to call what she does farming. But if you added up all the quarter-acre lots scattered among the high-rises of the south-central Bronx, then collectively they would qualify as a farm, making Karen one of the matriarchs of urban farming.[2]

FEEDING BODY AND MIND: THE GARDEN OF HAPPINESS

Karen and I drove northward through the Bronx.

"When people ask what I do, I tell them I am an urban farmer," she said. (She pronounced it "fahma," her Bronx accent having little use for most r's.) "I've got chickens. I grow food and feed people, body, and mind."

We were passing an endless succession of 99-cent stores, fast-food marquees, liquor shops, bodegas, and defunct commercial spaces. I marveled at the fact that this was Karen's agricultural zone. We began to discuss the research linking urban gardening with community health. Karen was aware of all this data and found it inspirational.

"I see urban farming as preventive medicine," she said, adding that no borough in New York City needed that prevention more than the Bronx. "We have by far the highest rates of diabetes and heart disease in the whole city. Something is wrong! If we look at our history, our parents were not popping pills for these diseases."

Given the enormity of the problem, I wanted to ask Karen how annexing scraps of land and throwing in seeds made a difference. But at that moment, we turned onto a quieter street of attached one-story brick houses.

This was Karen's own Crotona neighborhood. Each house had a fenced-in postage-stamp front yard with a wrought-iron gate.

"That's my house," she said, as we parked in front of a house in the middle of the row. It was easy to spot because of the greenery.

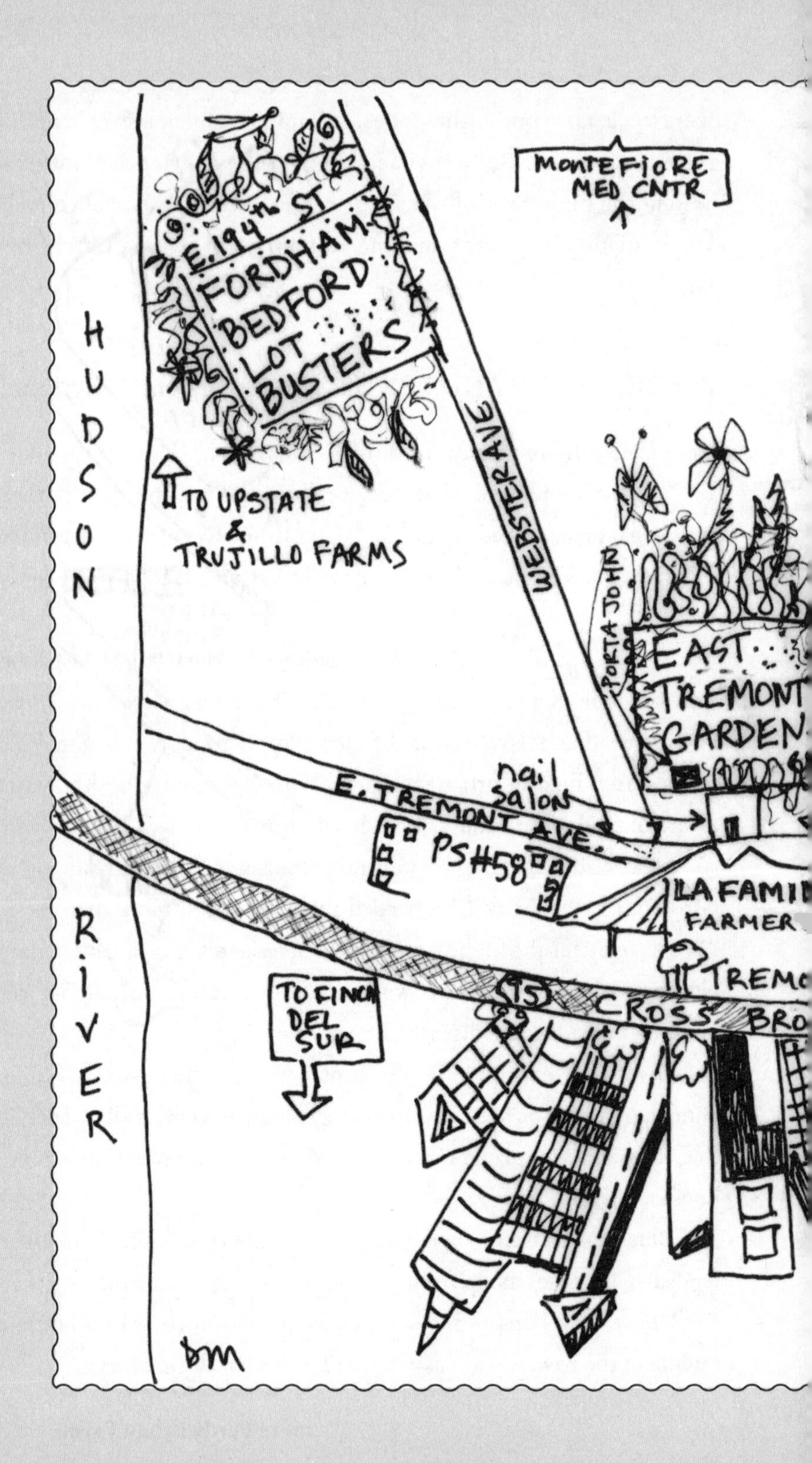
MONTEFIORE MED CNTR
E. 194th ST
FORDHAM BEDFORD LOT BUSTERS
HUDSON
WEBSTER AVE
TO UPSTATE & TRUJILLO FARMS
PORTA JOHN
EAST TREMONT GARDEN
nail salon
E. TREMONT AVE.
PS#58
LA FAMI
FARMER
TREMO
RIVER
15
CROSS BRO
TO FINCA DEL SUR
bm

La Familia Verde Urban Farms

"And that's the Garden of Happiness." Karen pointed across the street to a hurricane fence that housed her very first farmlet. Through the wire mesh I could see more vegetable beds, and a gaggle of red hens. "Just walking in there makes you happy."

We crossed the street, and Karen produced a custodian-size key ring. A plaque affixed to the fence gave a brief history of the creation of the garden. It mentioned Karen as one of its founding members, and explained that the garden was permanently assigned to the Department of Parks in 1998 and designated for farming in perpetuity.

"Having this guarantee is important," Karen said. "Otherwise, as soon as the economy changes these places become prime targets for developers."

Stepping inside, I immediately got that same valium-like feeling that I'd experienced in La Finca earlier that day: My pulse slowed, my joints released, and I could breathe a little deeper.

Although there are now dozens of community gardens throughout the Bronx, the Garden of Happiness is one of five gardens that collectively make up the mid-Bronx's La Familia Verde, or what Karen refers to as "our neighborhood farm." Individuals and families from the neighborhood tend their own plots, but each week they donate a percentage of their yield to the Mid-Bronx Farmers' Market. The proceeds go to the La Familia Verde organization.

"That's generous of them," I commented, impressed that the farmers give up a portion of their harvest.

"Well, when tomatoes are 'on their on', what you gonna do?" Karen answered. "Waste them? They're not even going to be missed from their plots. For us, it's not about the money."

The sunshine, the apple trees, and the colorful sheds made for a lovely scene. But I couldn't help but revisit the question I'd wanted to ask Karen earlier that day. How could this plot of land, as picturesque as it was, make a dent in the health of the surrounding community? The collards were on their last legs, as were the tomatoes and peppers, and the hens that followed us as we toured the garden had not laid eggs in days. Surely, when the plants were at their peak, plenty of valuable nutrition was being harvested from this little space, but in early November, I saw barely enough food to fill a few farm baskets. There would be a good seven months of fallow before this garden would start to produce once again. Also, other than the chickens, the cat, Karen, and Omar, I hadn't seen a single living soul in La Finca or Happiness. Where was the health-giving effect, and who was benefiting?

Just then, I noticed a white pedestal at the edge of the garden, standing alone under a mature apple tree. As I drew closer, I could see that it held up two statues flanked by potted geraniums. Closer still, I realized that the statues were of the Virgin Mary; the bigger Virgin, her robes painted a Caribbean blue, gazed down protectively over her smaller twin.

"That's our 9/11 shrine," Karen told me. "We put it up soon after that day."

We sat down on a nearby bench and gazed silently at the statues. The leaves in the apple tree danced overhead, and the hens pecked at the earth near our feet. The ground in this area was well worn, suggesting that over the years many people sat in this very spot. I had experienced 9/11 from three thousand miles away and did not know any of the victims, but here was a neighborhood that was less than thirteen miles from Ground Zero, with residents who were likely to have had a personal connection to the tragedy.

Near our bench in a central clearing stood picnic tables, an outdoor wash sink, and an oil barrel refashioned as a barbecue. It was surrounded by garden sheds painted in vibrant colors: fire engine red, a blue that matched the robe of the Virgin Mary. Party lights were strung through the surrounding trees. On warm nights, Karen told me, the garden is buzzing. People of all ages congregate to chat, play cards, cook, and tell stories. "Our families don't go off into a room and eat and drink and not talk to each other."

Now, everywhere I looked in the Garden of Happiness I saw signs of life and of human connection; this was a place that a community gathered to grieve, laugh, labor, cook, play cards, dance, and farm. Was it simply by creating these bonds that La Familia Verde and other urban gardens were able to increase vegetable intake and improve the health of their residents? I wanted to find out.

MARKET DAY

The next Tuesday, at 5:30 in the morning, I was slumped in one of the hard seats of the Seventh Avenue express train, once again northward bound. My plan was to help harvest the La Familia produce and then spend the day selling it at the weekly Mid-Bronx Farmers' Market.

I got off the elevated steel platform at the West Farms Square exit. The stop's name reminded me that long ago all this concrete was farmland. Just as I reached East Tremont Avenue, Karen sped around the corner in her Honda, slowing down just enough to let me jump into the backseat. She had driven to Penn State and back the day before to give the keynote at the Pennsylvania Women's Agricultural Network Conference. But even on four hours' sleep, she was in a terrific mood.

"Good morning!" she sang out. It was still dark. "Are you ready to sell some vegetables? Victoria here already did all the La Familia harvesting, so you're too late for that."

I looked over the headrest and greeted Karen's neighbor and cofarmer, Victoria Cabrera. I couldn't believe that this tiny woman was singlehandedly responsible for the big crates of fresh-picked collards, kale, tomatillos, and bunches of herbs loaded into the back. Victoria didn't speak much English, and Karen's Spanish was rudimentary, but they were communicating just fine. She told Karen that in addition to picking all those vegetables she'd also made a couple dozen chicken tamales, which she planned to sell at the market alongside the produce.

In the predawn light we zigzagged back and forth across the Mid-Bronx, briefly stopping to pick up a half-dozen more crates of vegetables from a delivery truck bearing the name Troncillito Farms. Karen explained that when La Familia's harvest started to wane, she used community garden funds to buy additional produce from organic farms upstate.

"Now, when I round this corner, I want to see blue," Karen said, taking a final right turn onto La Fontaine Avenue. "That tells me everything will go smoothly today."

Sure enough, directly ahead on the sidewalk fronting Tremont Park, I saw a series of blue tents erected over long folding tables. Behind these tables sat Randy, a hired assistant, and Mrs. Butler, a spry octogenarian who volunteered at the market rain or shine.

We opened the boxes of produce and Karen showed how to arrange it on the display table. Randy grabbed a box of romaine and began to toss the heads into a pile.

"Be careful with that lettuce," Karen cautioned. "You need to handle them like they're a woman." Randy laughed but slowed down and arranged the greens more artfully.

As the sun came up people from the neighborhood appeared one by one to lend a hand at the farm stand. Each one had a unique connection to Karen and a different reason for being involved with La Familia Verde and each one gave me a new perspective on how an urban farm (and farmers' market) can offer a community some powerful medicine.

THE FARMSTAND FACTOR

First to arrive was Shirley Edwards. Shirley wasn't too into digging in the dirt, but she considered herself as much a part of La Familia as Karen or Victoria Cabrera, thanks to her job as market treasurer. She told me that getting involved with La Familia had brought all kinds of positive things into her life. She'd started to eat a lot more vegetables and subsequently lost weight. Now she felt more energetic. She'd also reconnected to her roots: Her parents were from Puerto Rico, but she'd never learned Spanish and knew little about her family's culture. By spending time with other market volunteers and customers, she'd become conversant in Spanish and had picked up some traditional Puerto Rican recipes.

On market day, one of Shirley's jobs is to collect all the food vouchers from customers, enter them into her big ledger, and make sure La Familia is reimbursed by the impressive number of agencies and organizations—local, state, and federal—that subsidize foods purchased at farmers' markets. She listed each program on a finger: "There's Wholesome Wave, EBT cards or food stamps, Health Bucks, the WIC program, farmers' market coupons, Senior Money. You name it."

Just then, a teenage couple appeared, pushing a stroller toward the stand. It was far too early for your average adolescent to be out of bed, but clearly the baby—now happily napping—had reset their clock. The father hung back, rocking the stroller to and fro, while the mother began to select produce from the table. She examined the frost-damaged skins of the tomatoes with some consternation.

"They're perfect for salsa," said Victoria in Spanish. She was standing nearby, arranging the cilantro, and I could see that she'd been keeping a maternal eye on the couple ever since they approached the market. Victoria told the mother that the tomatoes were from her plot and then she shared her favorite salsa recipe. The young mother listened intently, asking a few questions and nodding at the baby's father, who was also paying attention. Finally, the couple chose four of the better-looking tomatoes and put them aside. Next, they turned their attention to a basket of pearl-size champagne grapes. These looked perfect, so neither hesitated to add them to the pile. They selected two heads of romaine lettuce, one bunch of cilantro, a yellow onion (for the salsa), five Macintosh apples, a couple of beets, and a pound of potatoes, and then they handed their food vouchers to Victoria. Shirley came over and helped with the accounting. After the Health Dollars and the farmers' market coupon, they owed only $5. Not bad for all that bounty! The couple, looking pleased, loaded their purchases into the stroller and headed up the street.

"If they went to a fast-food place, like a lot of teen parents do," said Shirley, motioning distastefully toward the golden arches up the street, "they'd spend a lot more than that on junk. It's amazing how many people don't have their priorities straight. They get their nails done but don't have money for fresh food." She pointed to the nail salon right across from us. "It's forty dollars for extensions and then fifteen for maintenance every two weeks. Just look at what that could get you at this market!"

Based on what I had just seen, it was easy to imagine how a farmers' market might inspire more vegetable eating than a supermarket. In a regular market, WIC dollars could be spent on sugared cereal and processed cheese just as easily as they could be put toward a head of broccoli. But here in the Mid-Bronx Farmers' market, only produce was for sale, with no opportunity to buy the unhealthy items that tempt us from the aisles in a grocery store.

The La Familia Verde stand was also more accessible than your average supermarket. Located on the sidewalk in the heart of the mid-Bronx, one could casually shop here en route to the nearby bank, Tremont Park, the neighborhood school, or the Mary Mitchell community center. Studies of food purchasing patterns in urban neighborhoods show that people are more likely to buy vegetables if they can do it in passing. A meaningful social exchange, like the one I just witnessed between Victoria and the young mother, makes the transaction even more appealing.

DIGGING FOR LONGEVITY

A pair of sturdy-looking, spectacled women appeared under the tent. They seemed no older than sixty-five, and I was shocked to find out they were both in their eighties. They greeted Mrs. Butler, who was bunching cilantro.

"Mrs. Nembly! Mrs. Weaver!" Karen cried. "Now all three of my queens are here to help me sell vegetables!" She extended her arms toward all three women as if to give them a collective hug. They giggled.

Mrs. Nembly and Mrs. Weaver, both born in Jamaica, still had their island accents. They addressed me and everyone else under the tent as "my dear." Mrs. Butler was originally from South Carolina, and everyone was "hon" to her. All three of the queens grew up on farms, and although they'd spent decades in the Bronx, each told me how much she loved her La Familia farming plot.

"And here is the king!" Karen announced. The three queens and everyone else turned their attention to a tall, dapper man with a salt-and-pepper mustache and a beret jauntily cocked to one side. This was Mr. Alexander, a truly indispensable member of the La Familia market team since, as treasurer for the nearby Tremont Garden, he held the keys to the only accessible port-a-john. Mrs. Butler leaned over and told me that he also happened to be a very good farmer.

Upon hearing that I was there to learn about La Familia Verde, Mr. Alexander asked if I wanted to see Tremont Garden, and I happily took him up on his offer. We crossed the street, leaving the three queens, Shirley, and Karen to serve the customers. Like the queens, Mr. Alexander had grown up on a farm, in South Hill, Virginia. He told me he'd left at eighteen to go to college and had lived in the Bronx for more than fifty years.

"I worked so hard to get off the farm, and promised myself I'd never go back," he said, opening the garden gate. "But here it became interesting again. On a farm you have to work from sunrise to sunset. Here you get out and mingle and give your opinion to those who haven't farmed. You work at your own pace."

The research I had collected suggests that while urban farming offers preventive medicine for people of all ages, it is especially valuable for older adults. Gardeners over fifty were less likely to fall, be depressed, or develop dementia than nongardeners. Watching Mr. Alexander tend to his plot, it was easy to see how this was the case. He lifted, pulled, dug, and pushed, sometimes balancing on one foot like a dancer. These were all activities that helped him maintain equilibrium and strength, and stay impressively limber. At the same time, he was making a thousand small decisions, this flower does best here, that squash over there. . . . Exactly how this brain work wards off dementia is not clear, but it seems to be highly effective. (Studies show that pursuing any kind of engaging leisure activity from golf to chess might have a similar benefit when it come to preserving cognition.)

Two older men sat at a table near Mr. Alexander's plot, their heads bent intently over a checker board. Two women, their contemporaries, sat on a bench chatting.

"They sit here and get to talking and sometimes I have to escape back to my office," Mr. Alexander groused. But I could see that he was just joking. For some people, growing old can be isolating, but from what I observed (and what the research shows) urban gardening offers plenty of opportunity for life-long connection. The key is to just keep gardening. One study I came across showed that when healthy, well older people gave up gardening, their health declined within a twelve-month period.[3]

CONTAGIOUS VEGETABLE EATING

A few school-age boys entered the garden. School was closed for election day and they wanted to help out. According to Mr. Alexander, kids come through the garden for all kinds of reasons: to visit grandparents, to play with the kittens, and to volunteer in an after-school farming program. Sure, sometimes they helped themselves to a couple vegetables—but that's what we all want, right?

Each boy took up a hoe and began to weed. I noticed one pause, pick something off a vine, and munch. Watching him reminded me of an interaction I had witnessed an hour earlier at the market, when a young woman, three children in tow, had approached the farm stand. She had seemed focused on the beets.

"Yeah, you want them beets that we picked this morning." Karen had crooned. The woman had nodded enthusiastically. "Just sautée the tops with onion and garlic. Enjoy, my dear."

The woman had taken the beets and had led her children away, each one crunching an apple from the Garden of Happiness, a gift from Karen. Karen had turned to me and nodded. "That was good."

The look of sheer pleasure registered by the apple-munching children and by this boy hoeing the Tremont plot made me think I was witnessing an "infectious" moment. For these children, the sensual experience of eating an extra tasty piece of produce could translate into a lifetime of more vegetable eating, and the act of growing that produce only increases the likelihood that this will happen. One study that took place in the south Bronx and central Harlem, showed that children who ate from local farm stands were more likely to eat fruits and vegetables with every meal. Participating in a school or community gardening project had an equally positive effect, especially when it came to developing a child's appreciation for "difficult" vegetables, such as spinach, beets, and squash. These studies also showed that adults increase their vegetable intake by virtue of their contact with these young gardeners and vegetable lovers. Just like a virus spreads throughout a community, so does the tendency to eat fresh fruits and vegetables. Now here's an epidemic that any public health worker would celebrate!

URBAN FARMING: A HOTHOUSE FOR COMMUNITY ENTERPRISE

Back at the market, the early morning rush was over, and everyone under the tent was discussing the local candidates on the ballot. It was a midterm election, so there wasn't too much to say. The conversation turned to Michael Jackson; the doctor who was caring for him at the time of his death had just been found guilty of involuntary manslaughter.

"Poor Michael," Karen said, with the kind of compassion one normally reserves for a friend or family member. "He was surrounded by people who lied to him. His nose was falling off, and no one told him."

Kevin, Karen's cousin, appeared under the tent, carrying a cake tin under each arm. Having been laid off from his job on Wall Street, he decided to follow his real passion and become a baker. At Karen's invitation, he'd recently started selling his wares at the La Familia Verde stand and now had a lineup of regular customers. I realized that the farming network was an incubator for a variety of small-business ventures, including Victoria's tamales and Kevin's cakes.

"I hate to see people defeated by the system," Karen said, rearranging tomatoes into a daring pyramid in order to make them look a little more appetizing. She explained that something that had started as a set of gardens was now a much more complex organization, offering community members a stepping-stone to new opportunities.

Economic stimulus was not a factor I had considered when I contemplated the public-health benefits of community gardening and yet over and over poverty and lack of opportunity are independently correlated with chronic disease and decreased life expectancy. Offering a sense of purpose and developing new revenue streams was yet another way that La Familia was fostering community wellness.

MORE VEGETABLES, LESS CRIME—WHAT'S THE LINK?

"Get your freeesh vegetables. We grow 'em, you eat 'em," Karen called out to shoppers, using her hands as a bullhorn.

Several dozen preschoolers led by a teacher appeared at the end of La Fontaine Avenue. In their brightly colored parkas and backpacks, they looked like a procession of chirping parrots. When they crossed the street and passed the farm stand, Karen waved at them and started to sing: "Remember to eat your vegetables!" They waved back, and one of the teachers smiled. Clearly they all knew Karen.

The children walked in pairs under the Tremont Park archway, headed for the playground. From the sidewalk I could see that the park was lovely, with gravel paths, perfectly manicured lawns, and maple trees displaying brilliant fall colors. Karen told me how the park had changed as a result of La Familia Verde.

"Eight years ago, this place was filled with drug dealers," she said. "Then the market started coming here, and next thing you know the Department of Parks is taking care, the police stop by more often, and the dealers disappeared."

Decreased crime rates! There was another health advantage of community gardening.

This talk about crime gave me an idea. Digging through my backpack, I pulled out a series of articles by the sociologist Robert Sampson, studies that Jill Litt had referenced in her papers. I found the titles intriguing and had skimmed them several days earlier on my flight to New York. At the time, I had found them discouragingly dense and overloaded with sociology jargon. But now, I reread them with new interest. According to these papers, Karen was right. The farmers' market had lowered the crime rate. But the authors' explanation for this was much more complicated than the mere fact that the market was adjacent to the park.

COULD IT ALL BE COLLECTIVE EFFICACY?

Sampson and his colleagues surveyed thousands of Chicago residents to understand why crime rates varied so much from neighborhood to neighborhood, even within the same socioeconomic groups: some poor communities experienced virtually no violent crime, while others were plagued by it. Similarly, although overall crime rates were lower within wealthier neighborhoods, some affluent places consistently saw more assaults and burglaries than others. They concluded that what made the biggest difference in crime rates was not economic status but something the authors referred to as "collective efficacy"—a community-wide belief that the members of the community could make a difference by working together.

A neighborhood with a high degree of collective efficacy is more likely to have a volunteer community crime watch, to report something suspicious to the police, to tell a parent if a child is skipping school, or to form a "safe routes to school" committee. Collective efficacy is also linked to many other positive health outcomes, such as higher self-rated health and fewer days of disability.

So what increases collective efficacy? Pretty much everything I had witnessed in the gardens and at the farmers' market. The multigenerational collaboration, the storytelling, the shared (delicious) meals, the incubation of businesses, the political discussions, the physical labor in the sunshine, the kindness to neighbors, the volunteerism at the market, the connection to family tradition, the sense of self-sufficiency, the interface with local agencies, the youth involvement—these were all sparks of collective efficacy that La Familia Verde had ignited in the once-blighted mid-Bronx. And what had sparked La Familia Verde? A handful of instigators like Karen Washington. Who would have thought that building a few vegetable beds in a vacant lot might eventually lead to such varied outcomes as less crime, less dementia, lower body mass, and more vegetable eating? It was amazing to discover that urban gardening could offer such effective medicine and yet, until recently, I had not prescribed it to a single patient.

TO BE HEALTHY, YOU HAVE TO LOVE THE PLACE WHERE YOU LIVE

Later, I called Jill Litt in Denver. The enthusiasm in her voice matched her sunny photo on the University of Colorado website. Litt told me that she had a Ph.D. in environmental health from Johns Hopkins and that her initial interest was in urban toxic waste sites and cleanup. She'd conducted many interviews in the low-income neighborhoods of Baltimore for her dissertation.

"I was talking to neighbors about the dangers of industrial properties," she said, "and they were much more focused on food and safety from crime."

Litt recalled that her career changed course when a colleague named Patricia Hynes—author of *A Patch of Eden*, a book about inner-city gardeners—invited her to a picnic celebration at the Dudley Street Project in Boston. "Families had converted vacant lots into gardens, they were taking pride in their work, there were people of all ages. And, oh, the fresh food. . . ." She paused, clearly enjoying her memory of that event. "That was the day I stopped doing traditional risk science research and reimagined my work to focus on land reuse and health."

Litt told me that she has become so passionate about urban gardening that she's transformed the front yards on her block into a quasi-community farm. Then I shared my experiences from the Bronx and we discussed how "collective efficacy" was an important mediating factor in all of the health outcomes I had witnessed there.

"But you know," she said, "at the end of the day I'm not even sure whether it needs to be gardening."

"Really?!" I said.

"Yeah, I suppose if you were a community of bird watchers, you might also have all these positive health benefits. What we've found in our studies is that neighborhood beauty is strongly related to health-promoting processes such as collective efficacy, which, in turn, relates to one's self-reports of health. If you don't think your surroundings are beautiful and worth engaging with on a daily basis, then you won't go outside, move, connect with others, or be active. The bottom line is that it's all about aesthetics."

I thought of Tremont Garden and the boys and seniors seeking refuge there, the people who had erected the 9/11 shrine, and the pride of place expressed to me by most of the people I'd met that day at the market. One thing they had in common was that they all saw beauty in their neighborhood. But unlike the beauty found by bird watchers, this beauty came from something they had created: the gardens of La Familia Verde. And it was this sense of beauty that motivated them to get out and move and till the soil and communicate and grow vegetables . . . vegetables that, in turn, would make them even healthier.

Now that is a sustainable cycle.

GARDEN OF PLENTY

The market ended at three o'clock, and I rode with Karen back to her house. She got out of the car and handed me a bag of vegetable stalks and other trimmings collected over the course of the day.

"Bring these to those chickens and talk to them. Tell them: You get fresh vegetables every week—open them legs."

Karen took off down the sidewalk carrying a leftover bag of vegetables and yelling, "Cheryl!" She was bringing them to a neighbor.

I sat back on the bench near the 9/11 shrine and rested, watching the chickens enjoy the scraps. Other people were there now working their plots, but still the garden felt like a refuge compared to the hubbub of the market. I looked at the garden beds with fresh eyes; I now understood how small patches of vegetables could impact the health of a community far beyond their unit dose of vitamin C or beta-carotene.

The epazote might be used just a few sprigs at a time, but it demanded to be mixed with a big pot of hardy, high-fiber, high-nutrient beans. Tomatillos, no matter how few, were begging to be slow-cooked with chicken, fiery chiles, and cilantro, and served with tortillas made from the maize that grew along the fence. Collards and yams this fresh needed little else, which meant that they weren't soaked in salt, preservatives, and unhealthy oils like the flavorless kind sold in cans. And once you started to prepare collards and yams at home, you were inspired to hand make the rest of your meal rather than get it from a fast-food chain. . . . How about that patch of yerba buena? A mood elevator, it could be plucked and steeped, offering instant herbal happiness to anyone in the garden. I wondered if this particular herb originated in Cuba, Puerto Rico, or perhaps the Dominican Republic, and thought of all the seeds from faraway places, carried here by this generation or the ones that came before. The product of those seeds offered La Familia gardeners a special identity and a connection to a traditional diet that was much healthier than anything they might encounter in a nearby bodega, convenience store, or supermarket. All this preventive medicine was here for the picking.

SOUNDVIEW: FROM GARDEN BED TO SICK BED

I found Karen in her sunny kitchen. She had changed out of her grubby farming clothes into fresh-pressed khakis and a clean shirt. Putting on spectacles, she fished in her bag for an ID badge, which she hung around her neck.

With less than four hours' sleep in the last twenty-four, Karen was about to start her day job as an in-home physical therapist for a local hospital.

This time, we headed east toward the Soundview area of the Bronx. The neighborhoods looked more industrial, less tight-knit than her Crotona area, and I didn't see any gardens. At each stop, Karen led her housebound patient through a series of gentle exercises, all the while discussing foods and recipes. One patient, Mrs. P, was riddled with pain from cancer and advanced rheumatoid arthritis. She said that her pain medication had made her constipated and so Karen gave her a list of foods that would help with constipation. Pears were on the list, and she promised to bring some the next week from the market.

Our final stop was the Castle Hill Projects, a maze of brick high-rises surrounded by concrete. A young man in stained boxers greeted us at the door of a fourth-floor apartment, looking as if he hadn't been outside in months. In an airless back room with peeling plaster walls, we found Mr. R, still wearing his hospital pants and sitting on a sagging mattress in front of the TV. Mr. R was forty-nine years old but looked ancient. He had end-stage renal disease from diabetes and had recently had cardiac bypass surgery. Unfortunately, his surgical wound became infected with MRSA, a resistant form of staph, and now he was bedridden from pain and fatigue.

There was so much stuff piled on Mr. R's floor that there was barely any room to move. But still, he and Karen managed to do some gentle exercises using the bed and the television as supports. As Mr. R slowly flexed his biceps and raised his legs, they chatted about sports, both lamenting a shortage of real heavyweight boxing champions, and for a moment he seemed to forget his dire circumstances. On the way out, Karen deposited a bag of produce in the kitchen. It seemed as if everyone got a piece of La Familia Verde.

Out on the street, Karen told me about her dream for a mobile market.

"It would be kind of like the ice cream truck, and it would bring vegetables to everyone here." She looked around at the treeless landscape and the concrete. "Look, there's nothing here." The fact that Hunt's Point, the central produce terminal for all of New York City, was just across the sound—less than two miles as the crow flies—made it all the more poignant that Castle Hill was such an extreme food desert.

"Food is political," she said, sighing. "The quality of food you get is directly proportionate to your political clout. And these people, they're eating the dregs."

As we drove back to the west side of the Bronx, I thought about some of the prematurely sick and disabled patients I had taken care of over the years. Many of them lived in San Francisco neighborhoods considered to be food deserts: Visitacion Valley and the Bayview. As in the mid-Bronx, residents of these areas have a life expectancy that is fifteen to twenty years less than that of the surrounding, more affluent communities. Of course, better access to primary care, better transportation, better housing, better schools, better jobs, more parks, and maybe even more supermarkets are needed to reverse these statistics. But recently, I've begun to see something in these neighborhoods that might be an equally important intervention: vacant lots slowly morphing into gardens with all the food-oriented activity that inevitably follows—farmers' markets, mobile markets, food swaps, gardening classes, cooking classes, and productive home gardens. The San Francisco fog might prevent a terrific bounty, but (as we know) that is beside the point: The mere presence of these oases will help improve the health of these neighborhoods. Karen seemed to be reading my mind. She told me that almost everyone she touched as a PT could have had a very different fate if she had been able to touch them sooner as a farmer. She started to talk about her plans to retire.

"Thirty-three more months, baby. Then I can do this full-time." She was fifty-seven, and she could retire at sixty.

We were back at 130th and Grand Concourse, the exact spot where I began my Bronx farming adventure. Karen parked, and we got on the train headed into Manhattan. I planned to go straight to bed, but Karen was headed somewhere in the Village to give a radio interview.

A PRESCRIPTION FOR HEALTHY PLACE-MAKING

Put It Up Front

As Karen Washington discovered, gardening in your front yard can inspire your neighbors to do the same. Jill Litt views front yards as the next frontier of community food and suggests replacing your front lawn and ornamental shrubs, sun and space permitting, with annual flowers and food crops. Front-yard gardeners, like community gardeners, are much more likely to interact with their neighbors, and the gardens themselves send the message that this is a vibrant and productive neighborhood. Unfortunately, there are stories from around the country of neighborhood associations and city boards penalizing front-yard gardeners for creating nonconforming, visually unappealing public spaces. Perhaps when the true health benefits of these patches are understood, shrubs and lawns will be considered unsightly and tomato vines and carrot stalks will become the prevailing aesthetic.

Small Is Beautiful

Community gardens and public green spaces need not be large, they just need to be accessible, protected, and enticing. Even a well-designed, well-kept pocket park with a comfortable bench serves an important role in maintaining community health, as it draws people out of their homes to move, interact with others, and enjoy their neighborhood.

Bring It into the Schools (and Elsewhere)

Research shows that school gardens and "edible education"—using gardens to teach children a whole range of skills, including cooking, language arts, math, and science—are effective ways to grow a new generation of vegetable eaters . . . and gardeners. I have volunteered in several of the Berkeley Unified School Districts gardens (where both my children are enrolled) and have seen firsthand how spending time playing in the dirt inspires children to be outside and sample a strange vegetable they might reject at home. As I mentioned earlier, health benefits from an edible education extend beyond the schoolyard, with parents and other family members of grade school gardeners reporting an increased vegetable intake. By the way, nursing homes, local businesses, religious gathering places, and community-based hospitals and clinics offer other opportunities to create social connection and expose a neighborhood to gardening and all its health-giving effects.

(Sharing) Food as Medicine

Host a vegetable swap or a pop-up picnic—sharing food is a terrific way to get to know your neighbors and to build "community efficacy," which we now understand as an important mediator in community wellness.

Start Simple

Begin by making it a goal to get dirty and grow some things with your neighbors. This simple act can be an important catalyst for improving the health of your community. As Jill Litt explained to me: "Environmental change alone is not enough to combat obesity and rising rates of chronic disease. Strong social organizations are also needed for populations to connect with these changes. Gardens are special because they are a local environmental change with a strong social organization." Yes, and strong social organizations begin with you.

What an Aromatic Herb Farmer Taught Me About Sustainable Beauty

For the first time in history, the goal of medical intervention is not restitutio ad integrum or restitutio ad optimum, but rather the . . . enhancement of the human body.

—Wolfgang Harth, Kurt Seikowski, and Barbara Hermes, "Lifestyle Drugs in Old Age" (2008)

GREEN WITCH and herb farmer Annie Harman cuts an impressive figure even with her two feet planted firmly on the ground. Standing high on the dais of the distilling shed, her silver hair one with a great cloud of steam, her long, bony arm thrust deep into the belly of a gleaming copper still, she looked otherworldly. Perhaps this vision was somewhat heightened by the eighty pounds of freshly harvested chamomile we'd distilled that morning. In the warm June air, the herb's sweet molecules filled the shed and saturated my brain, leaving me in a sublime, slack-jawed reverie. I stood there passively while Annie performed one of the last steps in the distillation process—draining her Dr. Seuss-like alembic stills and throwing the thoroughly stewed herbs into the bucket of her front-loader. From there the plant material would be composted and returned to the soil. Meanwhile, the results of our efforts—eight gallon-size jerry cans filled with an Easter-egg-blue floral water, or hydrosol—sat neatly arranged on the shed's floor.

"You see," Annie said, plunging her face into the steam to pull out an armful of chamomile, "I get a great facial every time I do this."

Earlier that day, I'd asked her how she managed to have such smooth and dewy skin—especially for a fifty-two-year-old—and this was part of the answer. But only one part. Over the next few days, my brown notebook would slowly fill with beauty tips—gems I'd traveled all the way to Morning Myst Farm, in the outer reaches of Washington State, to collect.

MISADVENTURES IN BEAUTYLAND

This last farm adventure was sparked by something highly personal and, in the scheme of things, shamefully trivial: a dark, caterpillar-shaped discoloration on my upper lip. I'm not sure if it appeared overnight or had been developing for some time, but I finally noticed it when on the eve of my forty-sixth birthday I was inspired to lean in a bit closer than usual to my bathroom mirror. Once seen, there was no ignoring the melasma, the medical term for skin that is hyper-pigmented, generally as a result of hormonal changes and too much sun. The mark, strikingly bracketed by the parentheses of my deepening smile lines, was all I saw whenever I caught my reflection.

My husband claimed he couldn't see it at all. But then, he was the guy who kept telling me I was gorgeous when, at the end of my first pregnancy, I was so swollen that my eyes were reduced to slits and bedroom slippers or extra-wide Tevas were the only footwear option. Only later, and under much duress, did he finally admit that I was less than comely during that third trimester. "Oh, that's nothing," said my friend Sharon when I casually asked her about my melasma. "I had really dark spots on my forehead and looked like Gorbachev." Sharon showed me a tube of lightening cream she'd been prescribed by her dermatologist, telling me that it had served her well. I looked at the ingredients: a dastardly trio of high-potency steroid, Retin A, and a bleaching agent called hydroquinone. At that moment, a cautionary inner voice reminded me of a promise I'd made to myself years earlier never to go down the slippery slope of medical cosmetic treatments. But how easy it is to be dogmatic when you have unblemished skin and nary a wrinkle! Surely a mustache-like stain was grounds for breaking that commitment. After all, it wasn't as if I'd just scheduled a facelift.

So hours later I was applying my own (self-prescribed) lightening cream, blithely ignoring the fact that the hydroquinone has been banned in Europe as a cosmetic because it causes cancer in rats. At first, things went well. I could see the darkness fading, replaced with a color that was a shade lighter than my normal skin, yet this seemed like an improvement. But then, as I neared Week Three of my regimen, I started to notice a new, less attractive reddish hue around my upper lip. Anxious to finish my treatment, I stifled my instincts to stop and pushed on for one more week.

Much to my horror, two days later I awoke to discover that the faint red had become beefy red, and the rash had spread to around my nose and to my chin, far from where I had been applying the cream. It looked like a bumpy crimson muzzle.

I took a picture of these latest developments and e-mailed it (along with lots of dramatic punctuation) to a dermatologist friend. Her response came back in minutes: perioral dermatitis, a grab-bag diagnosis that physicians apply to pimply rashes in the vicinity of the nose and mouth. And while the dermatology literature says that in most cases the cause is unknown, I knew for certain that mine was caused by vanity and an overly potent skin whitener.

COLUMBIA RIVER
Catnip
HERB CO. GROWERS
MORNING MYST
Annie & JOHN
COSMIC ENERGY ORGANICS
DM

Morning Myst

Any desperation I felt upon discovering the melasma paled in comparison to what I now felt. My kids said "ew" when I appeared at breakfast, and patients cringed as I drew near to examine them, clearly fearing that I might be contagious. Even my husband had a hard time pretending he couldn't see it. My dermatologist friend wrote, "The only thing that helps is antibiotics," so I promptly started myself on three weeks' worth of a slow-release tetracycline.

The drugs caused misery from day one. They gave me knife-sharp cramps after eating and a metallic taste in my mouth, and in combination with a glass of wine, they triggered one of the worst headaches of my life. But I kept going, because the eruption on my face seemed to get better with each passing day. Alas, twenty-four hours after I stopped the antibiotics, the rash and burning returned full force. Once again I sent the dermatologist an emergency e-mail.

"I hate to say it," she wrote back, "but you may need to take the antibiotics for longer, and there's still no guarantee it won't come back."

Suddenly I felt foolish. What was I thinking? I, who weigh all options before prescribing even a topical antibiotic or a low-dose steroid cream to a patient, had blithely put myself on weeks of a powerful steroid (plus retinoid and hydroquinone), followed by weeks of an oral antibiotic. There had to be another way.

I pulled out a reference book titled *Aromadermatology* written by Janetta Bensouilah, an acupuncturist/aromatherapist in the United Kingdom. I'd found it some time ago while browsing the Internet and was impressed by the caliber of scientific research on which she had based her recommendations. As the name implies, it's a book about treating skin conditions with plant essences or aromatherapy, and I've consulted it often to successfully treat a variety of skin problems—from eczema in children to acne in teens to psoriasis in adults. But somehow, with my own outbreak, it hadn't occurred to me to look there. Now, I turned to the chapter on perioral dermatitis and rosacea, two rashes that are considered to be close cousins in the world of dermatologic afflictions.

Bensouilah discusses how in both instances the stratum corneum (the skin's barrier layer) is injured and the normal balance of skin flora is disturbed, with an overgrowth of certain bacteria and yeast and a disappearance of others. (In my case, one can presume that the heavy-duty drugs were responsible, but other potential triggers include exposure to extremes in weather, illness, or caustic cosmetics.) She notes that the triangular zone between the nose and mouth is particularly vulnerable to these rashes, since the stratum corneum here is thinner than on other parts of the face. This area also happens to have a slightly higher pH and to run warmer than elsewhere on the face, a condition that promotes unwanted bacteria and water loss, and makes the skin more prone to injury. (This explains why my friend Sharon could tolerate the cream on her forehead while my applications in the lip zone wreaked such havoc.)

Bensouilah's discussion of skin architecture sounded very familiar. I returned to my bookshelf and pulled out *The Soul of Soil*, one of the important books that started me on my quest to learn from farmers. I randomly opened to a page on soil management and read:

> *Seedbeds are often created in a way that damages soil structure, resulting in erosion, compaction, and organic matter oxidation. For example, one common conventional method is fall moldboard plowing, spring dicing, and a final harrowing. The plow can create a hardpan with annual use, bury organic matter and living topsoil in an anaerobic zone, allow bare subsoil to erode by wind and water, and obstruct capillary action of water.*

The paragraph concludes with this: "The field may look good after the final harrowing, but not without great expense to soil tilth."

Since I'd spent time with Erick at Jubilee, I'd come to understand how closely we're connected to the earth. But this was the first time I'd fully appreciated how much skin, the biggest organ in our body, mirrors the soil in structure and function: Both skin and soil have three major layers and a top protective layer of dead cells (the humus or the stratum corneum); each harbors microorganisms in the upper layers; each is responsible for exchanging water, vitamins and minerals, and gasses with the outside world; and each is constantly producing new structures, whether hair, sebum, or blades of grass. Just as the health of the soil offers a barometer for the health of the farm, our skin—our most visible organ—serves as a gauge of our overall health. There was no denying that right there between the lower border of my nose and the tip of my chin lay a highly overplowed and overtaxed plot of land.

So, how to rebuild the stratum corneum? Bensouilah's advice includes avoiding really spicy foods, sugar, alcohol, and citrus that may cause direct irritation around the mouth or increase the temperature or pH of the skin. She suggests protecting this zone from the elements with a wide-brimmed hat, or a broad-spectrum sunscreen, and avoiding extreme cold, wind, and high heat. Finally, and perhaps most important, she cautions against using the caustic or allergenic chemicals, such as alpha-hydroxy acids, mineral oils, parabens, phthalates, SLS (sodium lauryl sulfate) and artificial fragrances, that are often found in cosmetics, shampoo, toothpastes, hairsprays, cleansers, moisturizers, and many fragrances. Sadly, I found at least one ingredient from this list in just about every product in my bathroom—including my "organic" toothpaste.

Then Bensouilah mentions hydrosols. She explains that these aqueous extracts are made from distilling whole plants and that one of their benefits, in addition to their cooling properties, is the astringent effect of the extract's essential oil. However, unlike hydroxy acid or alcohol-based cleansers, which can exacerbate or even cause dermatitis and rosacea by stripping the skin of its natural oils, hydrosols remove dirt and debris but leave the protective skin layer intact. Bensouilah's list of suitable hydrosols includes Roman chamomile (*Anthemis nobilis*), neroli (*Citrus aurantium*), rock rose (*Cistus ladaniferus*), lavender (*Lavandula angustifolia*), rose (*Rosa damascena*), and rose geranium (*Pelargonium graveolens*).

Hydrosol . . . that sounded vaguely familiar. I rifled through the bottom drawer of my bathroom cabinet and found a glass atomizer, a gift from a friend. It came with no instructions, so I used it occasionally as a pleasant, although disappointingly ephemeral, room freshener. The clear fluid inside smelled of citrus and roses and the ingredient list was reassuringly short: *Pelargonium graveolens* (rose geranium) hydrosol. Skeptical as I was, I decided I had nothing to lose and gave my face a generous misting.

To my amazement, the burning disappeared almost immediately—so quickly that I wondered whether rose geranium had anesthetic properties. I went about my day, occasionally taking breaks to spray my face and then seal it in with shea butter, a thick emollient with no ingredients from Bensouilah's danger list. Over the next few days, I boosted my diet with foods known to fight inflammation: onions and garlic, and three of my favorite spices—cloves, ginger, and turmeric. My skin swiftly improved—not completely back to baseline, but enough for me to feel confident that I was on the right track. Was it the hydrosols, the foods, or simply the fact that I had stopped doing caustic things to myself? Hard to know, but frankly, I didn't care.

At my local natural pharmacy, I looked for more hydrosols and I brought home a small but heady selection—lavender, chamomile, and more rose geranium—and continued to mist my way through the day. At some point, the rash completely disappeared, but I dared not stop my regimen for fear of a relapse. Also, I felt more fetching than I had in a long time—and calmer. People seemed to notice. Unsolicited, friends were telling me I looked great, and there seemed to be an uptick in the number of patients who casually asked for my healthy aging and beauty tips. Hand on the doorknob at the end of an appointment, they would say, "Oh by the way, what do you use on *your* face?" Was this all due to a hydrosol? I had no idea.

LESSONS FROM A CLINIQUE GIRL TURNED HOLISTIC AESTHETICIAN

Several weeks later, on a trip to San Diego, I decided to visit the woman behind evanhealy, the natural cosmetics company that bottled my first atomizer of hydrosol. The minute I laid eyes on the company's founder and namesake, my instincts told me that she was someone who would offer uncomplicated beauty advice.

Evan appeared to be in her mid to late fifties, and she glowed. Not with that lacquered look you get from overly stripped and tightened skin or glossy cosmetics, but with a soft light that seemed to come from within. As far as I could tell, she was still in possession of all her wrinkles and she wore hardly any makeup, save for the faintest of pink lipsticks that matched the pink of her silk kimono-style jacket. Her workshop/warehouse looked and smelled like a spa. In the main room, a handful of white-smocked employees sat at a long wooden table, listening to soothing music as they filled glass jars with creams and potions. They were working at a leisurely pace and seemed to take pleasure in handling the product. Evan introduced me to David, her husband and the head of marketing, whom she'd met years ago in a lavender field somewhere in the south of France. The couple led me into their sunny meeting room where the conference table was littered with dozens of atomizers identical to those now occupying my own bathroom counter, my desk, the hand-wash station in my exam room, and even my purse. I told Evan about my recent experience with her hydrosols, and she smiled empathetically.

"Between forty-six and fifty-six, the menopause years, we women are really on Mr. Toad's wild ride. I was in L.A. a couple of weeks ago, and I sat with clients in Venice, Santa Monica, and Beverly Hills. Now, this is ground zero for the beauty industry, but these are a subset of women seeking a more natural alternative to injections and surgeries. Even among this group, they had the same mantra: 'I don't understand what's going on with my skin.' I look at their skin and it's all buffed and polished—they've literally dissolved their epidermal layer with microderm products, alpha-hydroxy acids, and fruit enzymes."

Evan told me she'd started her cosmetics career years before as a "Clinique girl" at the I. Magnin department store. "I was selling products with all that junk," she said, referring to the range of ingredients used by the L.A. women. That's also what she used on her own face.

In the mid-1980s, her father retired from his corporate job and moved to San Diego to open one of the first natural food markets in the area. He asked Evan to join him to head the cosmetics department, and so began her "wholesale conversion from Clinique to plants." First she discovered Dr Haushka cosmetics from Germany, at the time a small, esoteric line. Haushka himself had studied under Rudolf Steiner (remember him from the Jubilee chapter?), and his products used botanicals that were grown and processed according to biodynamic principles and the rhythms of nature.

In 1988, Evan got her aesthetician's license and began treating clients. But as she gained more hands-on experience with the Haushka products, she noticed that they were not working for everyone. The line had exploded in popularity and she thought it was not being produced according to the high standards of the early years. Appreciating the value of freshly harvested botanicals, she began to incorporate locally grown plants into her beauty treatments. Eventually, she launched her own line of handmade, holistic cosmetic products. Her motto is "Farm to Face," and her label reads, EVANHEALY THE SKIN BREATHES.

"The more I learned about personalities of the plants and the growing process, the more I realized that graceful aging was a matter of the soil," Evan explained. "For example, like a wine, the lavender from 2010 had different notes than 2012. And these differences translate into different effects on the skin."

I asked for her top tips for healthy aging, and she offered me the same suggestions I'd heard often enough and frequently dispensed myself: eat well, exercise, get a good night's sleep, manage stress.

"But you know," she cautioned, "we have to get over our obsession with freshly minted body parts and let go of our white-knuckled grasp on youth. No one talks about beauty these days, they talk about anti-aging . . . anti-aging goes against nature. You can only hold your finger in the dike so long before the dam breaks. But feeling beautiful, you can do that at any age."

Then she got up and left the room, returning with another small diffuser.

"This is Immortelle [*Helichrysum*] hydrosol, a special one. Really quite rare." I misted my face: a little leather, a little tobacco, a little rose, a lot sexy. I was caught off guard by how intensely I was drawn to the smell. "I thought you'd like that one!" she said, laughing. "We live in such a complex world, and our skin needs not more complexity, but less. This plant doesn't strip your skin or interfere with its own natural functions. It offers you a simple beauty."

"But how does it work?" I asked, fighting through the intoxicating effect of the Immortelle and trying to get back to the hard science.

Thoughtful, she responded, "Well, I could say that the skin is a living organism and plants are living organisms. Or that we are a mediating influence between terra firma and the cosmos, born to receive and diffuse aromatic molecules. But of course, I don't think that answer will satisfy the doctor in you." She suggested that I visit the farmer who grows and distills her hydrosols, describing her as a walking herbal encyclopedia. That's when I learned about Annie Harman.

MORNING MYST

On my way north from Spokane, Washington, I cut through vast plains planted in wheat. It had been a mild June day when I left the airport, but here the crosswinds rattled my rental car. Occasionally the sea of grain was interrupted by a homestead—crouched behind a windscreen of poplar and fir.

Somewhere near the forty-mile mark, the road began to narrow and twist. I passed through a crack in the earth and entered into another world. Here homesteads and fields were nestled in clearings borrowed from the woods, and bends of the Columbia River could be spotted in the distance. In contrast to the windswept plains, this felt like a place of abundance, with plenty of water, game, and other natural resources to sustain local farmers. As Annie later put it, "If the world were to disappear tomorrow, these people would know how to survive."

Annie's farm is on a dirt road near the tiny town of Fruitland. Originally from Wyoming by way of coastal Washington, she moved here seven years ago, bringing little more than her impressive library of herbal plants, cuttings of her favorite herbs, and of course her alembic copper stills. The property belonged to her boyfriend, John, and Annie's friends decided that she must have been hopelessly in love to move to a trailer in the middle of nowhere. Meanwhile, the local farmers voiced their skepticism to John for having taken on this strange herb woman with giant metal contraptions.

"S'pose she'll winter out?" they asked him.

Not only did Annie survive, she thrived. The herb farming and distillation, which previously had been no more than a hobby supported by her day job, soon became her living. Her clients, including evanhealy and other high-end natural cosmetic companies, were so impressed with her hydrosols that they were pre-ordering them a harvest in advance. Soon the demand was too strong for her to meet alone, and she approached other farmers in the valley, proposing that they too grow aromatic herbs. Annie and her crazy stills had brought a welcomed boost to this economically depressed farm community.

My first night in Fruitland, Dick and Joan Roberson, an older couple involved in the herb enterprise, came to Annie's house for dinner. Before Annie moved to the area, they survived by growing alfalfa and garlic and by raising sheep for wool and meat. Dick was responsible for their herb venture. I asked him what he thought when Annie first came to him with her proposal, and he chuckled.

"Well, I thought it was all a bit strange."

Dick made it clear that he had no interest in cosmetics or, for that matter, in anything else that was only skin deep. What convinced him to plant an acre of aromatics was the per-pound price that Annie was offering to pay. (Her essential oils now fetch up to $300 per ounce and her hydrosols, at $120 per gallon wholesale, are more costly than many fine wines.) Although finances offered the initial incentive, Dick said that he and Joan have fallen under the spell of the herbs and would continue to grow them regardless of profit.

"You know when I really feel alive?" he asked, stretching his long arms out to either side. "It's when I walk through my peppermint field like this, with my fingers touching the tops of the plants. That smell. . . ."

I understood Dick perfectly. It was hard to find words to describe the effect of these aromatic plants. And yet it was words that I was looking for.

VIRIDITAS

The next morning Annie and I enjoyed our coffee on her deck. Below us, a big bend of the Columbia River swooped in to meet her fields. I told her about my experience with the rose geranium hydrosol and asked her why it had cleared up my skin so effectively. She answered by telling me about her own herbal education.

Even in middle school, years before she studied botany at Colorado State, Annie would pore through herbal pharmacopeias and concoct plant-based potions. She tried everything with botanicals: macerating them, adding them to creams and oils, stewing them in teas, cooking them in food, and making poultices. But there was always some dissatisfaction. As she described it:

"I felt that somehow I wasn't using plants optimally, and that there was a better way to access their true healing essence."

Then, about twenty years ago, she saw her first alembic. The Arabic word for "still," alembic generally describes any vessel that uses heat and boiling water to separate two substances. In Annie's case it refers to a copper still with an onion-domed lid—exactly like the top of a traditional Russian church—that she uses to create hydrosols.

"Seeing and smelling the rose hydrosol that came out of the still was life-changing," she explained. "It was a completely new way to enjoy the plants. To this day, thousands of gallons later, it still gives me a rush to see the first drops of distillate come over. Partly because of the mystery and magic and partly because of its connection to something so ancient." (Stills dating back four thousand years have been discovered in a perfume factory on the island of Cyprus.)

Annie read everything she could about distillation and then studied under Jeanne Rose, a renowned San Francisco herbalist who, by all accounts, is the originator of the term "hydrosol." Annie realized that while essential oils are derived from only fat-soluble molecules within the raw plant, each drop of hydrosol offers a perfect representation of the *entire* plant: water-soluble and fat-soluble. Annie made the analogy to a kaleidoscope, where each mirror is a tiny reflection of a larger image. She also discovered that aqueous hydrosols are the ideal suspension and pH to be readily absorbed and assimilated by our water-loving (hydrophilic), slightly acidic skin cells. In short, she saw hydrosols as the best way to access the healing properties of the whole plant.

"We are mainly water, and we crave water—we see it as something friendly," she explained.

"How about boiling the plant and using the water?" I persisted.

"You could take a handful of that fresh chamomile we distilled yesterday and steep it in boiling water for as long as you like, but I can tell you that the water will never turn the beautiful blue color of the distillate. We know from chemical assays that when that plant is steamed, the volatile oil Matricin is converted to Chamazulene, giving it totally different properties." She then rattled off a list of similar transformations that take place when other herbs undergo distillation.

"But don't get overly fixated on one chemical ingredient," Annie cautioned me. "If you do that, you're thinking with your drug industry cap on. You're trying to make a drug out of these herbs. But hydrosols are much more than their individual ingredients."

Annie had a good point. In a recent study involving witch hazel, researchers discovered that a distillate of the plant blocked elastase production in skin cells, elastase being a key enzyme in collagen breakdown and wrinkle formation. However, when these scientists studied hyaluronate, a witch hazel ingredient found in many high-priced commercial cosmetics and thought to be responsible for the plant's anti-creasing effect, they found that it had no impact on elastase. Similarly, hydrosols such as lemon balm are powerful antibiotics, even against bugs as daunting as Methicillin-Resistant Staph species. Yet isolated components of lemon balm are less able to eliminate infection.

Annie explained that when we "pharmaceuticalize" herbs and try to isolate a few active components, we run the risk of overlooking important ingredients whose concentrations seem too minute to be consequential. She gave the example of coumarin. Although it can barely be detected in lavender, lavender without coumarin loses its sedative effect.

Finally, she told me that "nurture," or environment, has as much to do with the healing effects of hydrosols as the plant's genetics—their genus and species. She left me on the deck and quickly came back with two vials of liquid.

"These are both from yarrow, or *Achillea millefolium*, and yet look at the difference."

I held them up to the light: One was straw-colored, while the other was deep blue. I pulled off the caps and smelled them, and the clear one seemed a little more turpentiney and pungent, the other more floral.

"That one," said Annie, pointing to the lighter-colored liquid, "was grown about one hundred miles east, deep in Hell's Canyon, where it's hot and dry. The other comes from my fields. The label is going to read ACHILLEA MILLEFOLIUM for both, but we both know that these two hydrosols are completely different."

I got her point. Unlike pharmaceuticals, herbs with the exact same label can have radically different ingredients, interacting with our cells in different ways. It seems that the power of hydrosols cannot be described by their chemical ingredients or even by the plant's genetics.

Annie Harmon's Hydrosol Face and Body Cream

This is a soothing night or day cream that is ideal for sensitive skin.

INGREDIENTS:

- 6 ounces organic carrier oil (such as, calendula-infused oil, jojoba, or almond oil, or a combination)
- 8 ounces organic rose geranium hydrosol
- ½ ounce beeswax, shaved into curls
- up to 1 dram (3.75 ml) organic essential oil (your preference; I use rose geranium and lemon)

EQUIPMENT:

blender
spatula
one pint-size or two half-pint-size canning jars with lids

INSTRUCTIONS:

1. Sterilize the blender, spatula, and jar by wiping them with Everclear or vodka.
2. In a double boiler, melt together the oil and beeswax. Let cool until a "skin" develops on top.

3. Meanwhile, put the bottle of hydrosol in a bowl of warm water. It must be warmer than room temperature in order to blend correctly. Once warm, mix the hydrosol into the heated carrier oil.
4. Turn the blender on low and slowly add a stream of the infused oil mixture. As it starts to thicken, turn up the speed. When the oil mixture has been incorporated, it will be white and creamy. If a small amount of hydrosol remains on top, just pour it off.
5. Fold in the essential oil with the sterilized spatula. Do not overblend.
6. Place the cream in the sterilized jars and store the cream in the refrigerator. Use it within three months. Since the cream is perishable, never place your fingers in the cream; use a clean applicator.

OTHER USES FOR HYDROSOLS:

- For a winter boost to your immunity, add one tablespoon organic lemon-thyme hydrosol to one liter of drinking water.
- For acne, spray lavender or rose-geranium hydrosol in your nostrils and on your face.
- For heartburn, add one teaspoon organic peppermint hydrosol to one cup of water.
- For winter's dry skin, spritz on your favorite hydrosol of the day (such as rose-geranium or lemon-verbena) after a shower.
- For a good night's rest, put two teaspoons of St. John's wort and lemon balm hydrosols in your bedtime cup of herbal tea.

In the midst of this involved science lesson, Annie the chemist retreated and the green witch took over. Suddenly she was channeling Saint Hildegard von Bingen, the twelfth-century visionary who is known for her writings on mysticism and herbal medicine. Annie explained that the German saint wrote about *viriditas*—literally, "green-ness" in Latin—using it to describe a state of physical or spiritual health.

"On an energetic level, I believe that hydrosols are the closest you can come to capturing the viriditas of the plant. And when we use them, that beauty and healing are transmitted to us." Of course, viriditas is not something that can be tested with a biochemical assay or gas chromatography/mass spectronomy, the standard tools used to identify specific chemicals in plants. But this didn't seem to deter Annie or her many clients, who regularly wrote her e-mails saying that they could feel the special energy in her products and were willing to pay top dollar for this unquantifiable attribute.

"So what gives your hydrosols that extra viriditas?" I asked, still wanting something I could measure.

"It's the entire process," Annie replied. "Saving the seed, tending the soil, sowing the seed, harvesting the plant, worrying over the distillate, and then loading the remains back onto the field." In other words, just as Cody Holmes, Erick Haakenson, Jeff Wheeler, and Karen Washington had taught me, the real answer was not in the details but in the whole.

BEAUTY TIPS FROM A MAKEUP MOGUL

Bobbi Brown, the owner and CEO of her eponymous makeup line, isn't familiar with the term "viriditas," but I'm quite certain that when she talks about "essence," she too is referring to that unquantifiable life force. She had e-mailed shortly before my trip to Fruitland because she'd seen a review of my last book and was interested in chatting. Curious, I proposed that we meet. I was hoping that I might get some valuable cosmetic tips from the woman whose impressive list of clients includes Michelle Obama, Meg Ryan, and Naomi Campbell.

"In my work with women, my goal is to help them bring out their true essence, the energy that shines out of each of them and makes them unique," she explained the minute we were seated in her favorite health food restaurant in lower Manhattan. It was a rainy March day, and I arrived to our meeting a frizzy, sticky mess. Either Bobbi was being polite or she appreciated the authenticity of my look, because the next thing she did was tell me I looked lovely. Regardless, I liked her immediately.

I asked Bobbi how she taps into a woman's essence, expecting that our conversation would quickly turn into a lesson on makeup application. But her answer surprised me.

"I think the biggest fix I can offer is to help set the reset button on negative self-talk. When women feel confident, that shines through. I call it 'pretty powerful.' " In private consultations, her intervention might include getting a woman to wash her face and take off all her makeup, or getting her to stop looking at herself in the mirror in the morning or checking her reflection in the distorting screen of her iPad (two activities that she said are major confidence bashers). I noticed that Bobbi herself did not seem to be made up. Her straight black hair was pulled back in a demure ponytail, and she was wearing dark-rimmed glasses and yoga pants.

"And then I always tell women that at a certain age they should stop worrying about their face and start worrying about their body." She continued, discussing her own growing interest in daily exercise and cooking fresh, locally sourced meals with her family. She told me that eating animal products and dairy sparingly, avoiding fluffy breads and processed grains, and switching to a vegetable-rich diet had done wonders for her skin and figure. I was intrigued that someone who worked in an industry that was all about quick fixes was so focused on subtle, long-term beauty solutions that take place on a cellular level.

Trying again to bring the conversation back to the superficial, I asked for some makeup pointers. "Sometimes the smallest things make a big difference—a thin black outline to the eyes, or a bit of lightness under them that's one shade up from your natural skin color, or a little pink to the lips. I believe women should wear just the right amount of makeup to make them look like themselves."

I persisted, asking her what she suggested for women my age, who were gathering more wrinkles by the day.

She set her fork down and leaned forward.

"I don't believe there's any product on the market that will get rid of a wrinkle. But I happen to think women look better as they get older." She told me she's drawn to things that are asymmetrical and not perfect. Even when choosing furniture, she prefers pieces that are a little chipped and worn. Then, pointing to her own regal nose, she told me that when she was twenty-one her mother told her she'd look prettier with a smaller nose. She responded, "I never said I didn't like my nose."

"I see so many women do things to themselves," continued Bobbi. "They might look perfect, but they lose that essence." I thought of my friend Carolyn Chang, a plastic surgeon in San Francisco who, interestingly enough, shares Bobbi's belief that imperfections are very often beautiful. She cautions her clients against having procedures that make them appear overly "done."

As we were getting up to go, she came around the table and inspected my face. I silently hoped that the eyeliner and mascara I'd amateurishly applied that morning had been washed away by the rain.

"You know, I love your essence," she said, again using that word. "Maybe I would suggest a dab of under-eye cover—what I call a minimal facelift." So there you have it. The CEO of a Fortune 500 cosmetics company, and all she wanted to sell me was a tiny pot of concealer! Or maybe this kind of minimalism was the secret to the success of her makeup line? I left our lunch, frizzy head held high. The next week a lifetime's supply of Natural-colored Bobbi Brown creamy concealer arrived at my doorstep in a UPS package. A small but very specific boost to my viriditas.

CRYSTALS IN THE CALENDULA FIELD

After finishing our coffee, Annie and I left her deck and drove out to a nearby hollow to visit Tracey and Russell Stringfellow. They are the farming couple who grow and distill Evan Healy's rose geranium hydrosol, the floral water spray responsible for launching me on this whole adventure.

Tracey is from San Francisco, and Russell grew up in the bohemian enclave of Vashon Island near Seattle, and they were familiar with herbal concoctions—in one form or another—long before they met Annie. But it was her distillation know-how that helped them convert their long-standing interest into a real source of income. When Annie and I arrived, they were up the hill in the geranium patch, tending to young plants.

"I have at least 150 varieties of wild herbs growing up here in these hills," said Tracey, pointing to the bushes upslope from her fields. She walked over to a nearby patch of weeds and picked a leaf from a grayish, bushy plant.

"Give that a sniff," she said, crushing the leaf gently between her hands and handing it to me. I breathed in deeply and immediately felt as if I'd stuck my head into a can of gasoline: The world looked fuzzy, and I noticed a twinge of nausea. Everyone laughed.

"That's wormwood," she said. "It grows wild here. We used to make extra money picking it and selling it to make absinthe liquor. But the smell made everyone too sick and crazy. All my pickers quit on me."

I followed Tracey, Russell, and Annie down the hill and past a calendula patch to a barn where I noticed a copper still, almost identical to Annie's. In the driveway we met up with their friend Sue Dahl, a nurse visiting from Hawaii. Sue had a gentle voice and a warm, guileless smile, both wonderful attributes for someone in the healing profession. She told me that after years of working with advanced medical technology in an intensive care unit setting, she was reorienting herself toward a less interventional, more energetic form of medicine. Much of this change was inspired by tending to her own mother in the weeks before her death. Sue's mother had suffered a stroke that left her paralyzed and mute, and yet in her last hours she seemed able to speak and move limbs that had long been frozen. Sue, with her background in science, felt for the first time that she was confronting a phenomenon that was inexplicable. Since then, she had trained in reiki and other forms of energy medicine, and she was now interested in healing through laying on of hands and through crystals.

I nodded politely at the mention of crystals, not sure what to say. Over the years I've ventured far from the confines of my orthodox biomedical training and have been amazed by studies showing that people respond to intangible interventions such as "distance healing." On many occasions, energy medicine has been of great therapeutic value for my patients, and at times I've personally felt an effect from hands-on energy work such as acupuncture, reiki, or chi gong. But crystals were another matter. It was hard for me to think of them as anything more than a mineral version of a Ouija board. I was glad when Tracey broke the silence and invited us into their farmhouse to sample some hydrosols.

The Stringfellows' living room looked like a diorama in a natural history museum. Atop the TV, a small catlike creature, stuffed in mid-strut, peered out from behind a clump of dried rushes, and nearby hung larger pelts from all sorts of animals. A mounted stag's head looked so lifelike that for a moment I thought he had just butted his five-point antlers through the wall to join our gathering. At the far side of the room I spotted more antlers and pelts and a large fish tank, and on the coffee table, amid bottles of rose geranium, calendula, yarrow, and lemon-thyme hydrosol, was a skull and four or five long bones—all unmistakably human. Russell must have read the panic on my face and laughed.

"Don't worry, we're not cannibals."

He explained that they had won a coffin in an auction at the local Odd Fellows' lodge—why they wanted a coffin, I am not sure—and inside were these bones, which, based on the other contents in the coffin, must have dated back to the late 1800s. Knowing that it's illegal to harbor human remains of any vintage, they immediately called the local sheriff, who'd not yet come to claim them.

First the druggy wormwood, then Sue's story from the edge of the grave and the taxidermy, and now this skeleton. . . . I was beginning to think that things could not get stranger—but then they did.

We all sat down around the coffee table, and Sue produced a pink crystal on a chain, which she held over each bottle of hydrosol, one by one. I watched, half-hypnotized as the pendant began a gentle clockwise swing above the atomizer.

"Now this is what's troubling me," she said, moving the pendant above the last bottle on the table. It was identical in size to the rest and contained lemon-thyme hydrosol. Amazingly, the crystal quivered for an instant, as if unsure what to do, and then began to swing counterclockwise, unlike all the others.

"The energy of this hydrosol is off," Sue said, shaking her head. She pulled the crystal away, then once again moved it over the bottle. I watched her hand to see if it was moving, but it seemed cemented in place. And once again, the crystal resumed its counterclockwise movement.

Annie didn't seem to be the least bit surprised by this finding. This batch of lemon thyme was for a specific client who insisted that she use dried herbs, which were cheaper but inferior to fresh ones. She guessed that this was what Sue's crystal was detecting. Judging by the nods, everyone else in the room found this explanation acceptable. Only I was having a hard time buying it.

I asked Sue if I could have her crystal. Bracing my elbow on the coffee table and focusing all my effort on keeping my hand still, I dangled it above the lemon thyme. It went in a counterclockwise direction. I changed position and held it over another bottle. Clockwise. I held it over some commercial sani-wipes. Counterclockwise. Everyone in the room was telling stories about encounters with crystals and the supernatural. Sue ran out to her car and came back with a beautiful rose quartz, also on a string, and gave it to me as a gift. Tracey, in turn, gave me a fresh bottle of rose geranium hydrosol.

I was almost too distracted to thank them. I had no idea what to think. Was my brain sending a signal to my hand to swing that crystal in a certain direction? I wanted data. A randomized trial. I wondered how I would set this up if I were in a lab.

I looked around the room. Everyone else had moved on, laughing, spraying hydrosols and sniffing them, sharing farming tips and local news, and having a wonderful time. At that moment, I realized I was focused on the wrong thing. It really didn't matter what made that crystal move: my hand, my brain, or a magnetic field. This ritual, for all of them, was one of many ways of measuring the viriditas or essence of their herbal craft, much in the same way that Cody might hold his buttercup-yellow raw milk up to the light to understand its beauty. What mattered most to these farmers was the care and consciousness that they poured into their herbs and distillates, apparent with or without a crystal. While the antibacterial volatiles in the hydrosol may have cured my rash, it was this care that made me feel lovely, that reset what Bobbi Brown called my "negative self-talk button."

Heading north with Annie to visit other farmers, I noticed a rainbow dancing across the dashboard. I wasn't sure if it was the rose quartz around my neck or the bottle of hydrosol in my hand that was catching and reflecting a ray of sunshine.

THE HUMAN DIFFUSER

After returning home, I began to dig through a stack of articles and books about the science of smell. One paper about human tears caught my attention. In this study, researchers asked three women to watch sad movies and, with the help of a hand mirror, collect their own tears in a glass jar.

They then recruited twenty-four heterosexual men in their twenties and asked them to smell (in no particular order) two unmarked jars, one holding the tears and the other containing a saline solution. All the men reported that the content of both jars was odorless. Next the researchers taped a cotton patch, alternately soaked with each liquid, under each man's nose and measured his reaction to a photo of an appealing woman. Although the men could detect no difference in smell, the tears consistently made them feel less excited. Lab measurements of sexual arousal, including skin temperature, testosterone level, and specialized brain MRIs, corroborated the men's stated preferences.

This study suggests that seemingly odorless tears can transmit messages that affect emotions, skin reactions, hormone levels, and even brain chemistry. As I reached further into the olfactory literature, I realized that my hydrosols could be producing effects very similar to tears, a fact which explains why they seemed—almost magically—to improve both my complexion and my psyche.

In his book, *Medical Aromatherapy*, Kurt Schnaubelt explains that nature in its ultimate simplicity uses the same substances and biosynthetic pathways over and over, a concept that evolutionary biologists and geneticists often refer to as "deep homology." He asserts that the mevalonic pathway, which produces aromatic volatiles, existed within the first single-cell life forms on the planet and served as the original form of inter-creature communication. This pathway was eventually transferred to multicellular plants to produce odiferous essential oils, and to humans and other animals to make pheromones—the hormones specifically designed for organism-to-organism communication.

Because of deep homology and the early universal reliance on olfactory messages, humans and other vertebrates have more active genes for smell reception than for any other sensory function. This fact is surprising given that smell is often perceived as playing a minor role in human communication. Deep homology also explains why a compound almost identical to the testosterone metabolite measured in those tear-sniffing men is also found in truffles and celery, and why clover produces a substance similar to human estrogen. Finally, it's the reason why Geraniol, the same volatile compound that helps honeybees mark attractive flowers or defend against an insect foe, can also attract us to a plum or peach, or to a person when used in the form of perfume.

I thought back to that moment in Evan Healy's conference room when I sprayed that remarkable Immortelle on my face and asked her how it worked. For some reason, I'd ignored the second part of her answer, writing it off as excessively New Age. But now, in light of what I'd just learned, it seemed important and rooted in a sophisticated understanding of neurochemistry. She'd said:

"We are a mediating influence between terra firma and the cosmos, born to receive and diffuse aromatic molecules."

Now it made sense that plants had rebalanced me when synthetic chemicals could not: Plants were, in essence, my distant cousins. Maybe there was something quantifiable about viriditas after all. Perhaps it represented the sum of chemical messages that we, or plants or any living thing, sent out into the cosmos. We just didn't know how to measure this very well . . . yet.

PLANT-HUMAN MISCOMMUNICATION

If there were a standard test for viriditas, most of the fruits and vegetables sold in our modern-day supermarket would receive a low score. This was what I came to realize after speaking with Beth Mitcham, a post-harvest researcher at the UC Davis school of agriculture. Mitcham explained that plant scientists were so focused on breeding for disease resistance, ease of transport, aesthetics, yield, and shelf life, that they bred the flavor out of produce. A plant's flavor is partly determined by its sugar and acid content, but the concentration and variety of volatiles (aromatic molecules) plays an even larger role. In short, our modern-day fruits and vegetables are highly deficient in the genes that code for the mevalonic pathway and all the aromatic messages it produces.

Mitcham contends that the standards set by the USDA have supported this trend of breeding for looks over taste. Government regulations dictate that no more than 10 percent of a fruit or vegetable can have a blemish—or what's called "inferior development"—if it's to make it onto the grocer's shelf. This means, according to Mitcham, that 20 to 30 percent of all produce fails to make the grade and is relegated to canned food, or animal feed, or is simply left to rot in the fields. Not only does the USDA fail to offer any standards for taste, but one might argue that the agency's exclusion of imperfect produce is a direct attack on fruit and vegetable palatability and nutritional value. This is because those tasty volatiles or secondary metabolites (which are also referred to as "antioxidants") are often produced in response to a stressful influence such as insects, molds, sunshine, or encroaching plants. These challenges, in moderation, help enhance a fruit or vegetable's flavor and nutritional punch.

While Mitcham sees everyone from agribusiness to government as part of the problem, she placed most of the responsibility squarely on us, the consumers.

"Research shows that food gets rejected at point of sale if it's bruised, but not if it's unripe or has poor flavor," she explained. "So there's no incentive on the part of the grower or the market to grow tasty produce. Some peaches might be fabulous, but if they have a little blemish you might pass them over. Customers buy with their eyes."

This is true. Most of us focus on the superficial, ignoring the universal (and ancient) language of aromatics when making our food choices. Certainly the same holds true when it comes to matters of physical beauty. We "buy with our eyes," seeking out the same uniformity and symmetry that we expect from a supermarket peach. Meanwhile, we have turned off that primordial sensory organ that clues us into more subtle, but ever more delicious, kinds of beauty. I thought about the older women who have been my own role models for lovely aging. The list includes Micki Colfax, mother to my doctor/farmer friend Grant; Tieraona Low Dog, the physician who taught me most of what I know about herbal medicine; Tanya Berry, Wendell's wife; and my mother, Susan Miller. I realized that the one thing these idealized older women have in common is that they spend a lot of time outdoors and communicate with weather, animals, plants, and soil on a regular basis. I thought again about the tear study and all the papers I'd read showing that pheromones and other aromas can regulate our internal processes and help shape our outlook on the world. These women (and the love partners who appreciate them) are still receiving and emitting these subtle—but potentially powerful—aromatic messages. Maybe this is the secret to sustainable aging! To be able to pick up these signals continuously—both among our own kind and among plants, other animals, and the ecosphere.[1]

CONSULT THE GENIUS OF THE PLACE

On my last day in Fruitland I awoke feeling more springy than I had in a while. I could only attribute it to the wonderful company, good food, fresh air, long evening walks in the woods, tomblike quiet at night, and a steady misting of plant hydrosols.

On my way to join Annie, I met her horse, Mystic. The mare is free to roam the property but seems perfectly happy staying close to home. I'd first noticed her a few days earlier when, at the height of the chamomile distillation, she poked her head into the shed to take in a muzzleful of vapor. Now, as I stroked her, I wondered if the luminous coat and extra-velvety chin were due to her regular herbal spa treatments.

I found Annie sitting at a Formica-topped work table looking like a chemist among her glass beakers and pipettes. She pulled out a plastic box filled with glass vials, each holding a sample of hydrosol or essential oil carefully marked with the plant's origin and the distillation date. Annie explained that these served as a databank to help track distillation methods and plants from season to season. In preparation for an upcoming presentation in Ireland at the Botanica 2012 conference, she was sending some samples to a chemistry lab for testing. The lab would check for certain volatiles and other plant metabolites such as flavonoids but would not measure tiny concentrations of ingredients that might play a vital role. And of course, there's no accepted assay for viriditas.

I noticed a small bottle labeled FOREST. It was a co-distillation of

Douglas fir and ponderosa pine, two tree types found on her property and throughout the Columbia River region.

"A co-distillation is very different than a blend, where you make the hydrosols separately and mix them. Here, they go in the still together and come out as something totally new. There's a synergy that's immeasurable—it's like capturing the whole forest."

"Forest" had a surprisingly delicate odor—like the woods after a rainstorm. I asked Annie how anyone would even think to throw fir and pine needles into a still. She handed me an old book published in 1903, its celery-colored linen cover worn thin at the edges. Embossed in gold on the front was the title: *The Still Room* by Mrs. Charles Roundell. Underneath was an image that looked remarkably like one of Annie's copper stills.

"This book was written at a time when every community had its own distiller and still," explained Annie. "And what you put in that still was literally what grew in your backyard. If you were living up here, you weren't importing ylang ylang from Polynesia, you were using something fragrant that grew outside your door. By the way, I believe that your local pharmacopoeia is also your best medicine."

This wasn't the first time I'd heard this idea. Annie, like Wendell Berry and many of the other farmers I'd spent time with, was calling on us to "consult the genius of the place."

Environmentally speaking, it made a lot of sense. Why should we burn the fuel to haul ylang ylang across the globe when rose geranium required no such input? And of course, there was the freshness factor. The rose geranium was likely to be much more potent, as research shows that most nutrients and other metabolites in herbs, fruits, and vegetables degrade over time.

But I sensed that our backyard pharmacies offered health advantages beyond freshness and carbon footprint—advantages having to do with the co-evolution of humans and plants and with their intracommunication. Years earlier I'd visited Copper Canyon in Mexico as part of my quest to understand why the Tarahumara, native Mexicans who live deep in the canyon, had such low rates of diabetes—a finding that was especially curious given that this group shares a common ancestry with the Arizona Pima, a tribe with exceptionally high rates of the disease. I'd learned that in addition to eating very little sugar and virtually no processed carbohydrates, the Tarahumara's cliff homes were encircled with more than three hundred species of wild foods that Mexican ethnobotanists had identified as playing a role in lowering blood sugar. In other words, they were surrounded by an antidiabetic pharmacy.

I flew back to the Bay Area the following day. As I pulled my suitcase along the sidewalk that leads to my house in Berkeley, I noticed for the first time that clumps of yarrow and geranium were growing through the cracks in the concrete, their flowers and leaves casting a lacy shadow. Every day, I step over these plants or crush them underfoot, without giving them the slightest notice. But now I dropped my bag, knelt down, and took in their smell. These were my yarrow and my geranium, and no doubt they had their own chemotype—their own unique chemical makeup that reflected my neighborhood. In my bag was a diagram for a stovetop still, a simple setup involving an enamel pot, a bowl, and a brick that Annie had given me so that I might make some of my own hydrosols. I gathered a clump of the plants and headed inside, determined to make my first homegrown beauty treatment. I pictured the glass atomizer filled with this new hydrosol. I would label it SIDEWALK.

Conclusion

Five "Ahas!" and One Suggestion

THE MOST satisfying moment in any process of inquiry is not when one confirms an already held suspicion, but when one is surprised by a new reality. My immersion in the world of farming offered me many such moments—whether they concerned farming, my own health, or the health of my family, patients, and community. While I would hope that you have had your own insights and inspirations along the way, I would like to leave you with five of my biggest "ahas!" and some final words about how you might transform them into action.

First, I now understand that the secret to good farming—a secret shared by all the farmers in this book—is to care more for the farm than for any product that farm may produce. (Hence Erick's favorite adage: "Feed the soil, not the plant.") If you had asked me before my internship at Jubilee about the hierarchy of goals on a vegetable farm, I would have told you that of course the greatest priority was producing vegetables—just as I presumed eggs would come first on an egg farm, calves on a ranch, wine in a vineyard, and pounds of distillate on an aromatic herb farm. These are the things with market value, the economic drivers that support a farm's existence. But all of these farmers are much more focused on the soil—both literally and metaphorically. They have come to see that in order for their farm to thrive they must treat it not only as a center for production but also as a rich ecosystem. In fact, they are more ecologists than farmers, because what occupies their attention, day in and day out, is the complex web of interactions among animals, plants, and nutrients.

Each of the patient stories also has a "plant" focus: weight and lab tests for Allie; eczema severity and ear infection frequency for Frankie; hemorrhoids for Mike; degree of esophageal cell dysplasia for Dava; or, for me, a facial rash. Health care providers, health systems, and patients chase these outcomes and make every effort to have them fall within normal ranges, even if doing so involves a medicine or procedure with negative consequences. For example, long-term use of the proton pump inhibitor prescribed to Dava to improve her Barrett's esophagitis is linked to osteoporosis, higher rates of pneumonia, and, when taken with a commonly used antiplatelet drug (clopidigrel), an increased risk of stroke.

Just as a whole-system approach was transformative for the farms, a shift in focus from "plant" to "soil" proved equally valuable for human health. Allie began to regain her energy when she stopped seeing herself as a collection of high to low labs in need of adjustment but as a member of a farm community; Frankie needed fewer medications and had fewer allergy symptoms when his parents connected his health to family dynamics and to his external environment; Mike's health status changed when he adjusted his relationships at home and at work; the Crotona neighborhood in the Bronx is tackling an epidemic of chronic diseases using a neighborhood-wide ecological approach; and I received my own lesson in complexity when I discovered that healthy aging is conferred from within, from one's loved ones, and by nature herself.

Although many of the positive changes described in this book involved dialing back the use of drugs and other technologies, my second epiphany is that eco-farmers are not by definition anti-chemical or anti-innovation.

In fact, they believe that modern science can play an important role as long as it preserves and complements the natural system. (As Aldo Leopold famously wrote: "If the biota, in the course of aeons, has built something we like but do not understand, then who but a fool would discard seemingly useless parts? To keep every cog and wheel is the first precaution of intelligent tinkering.") Take Cody of Rockin' H Ranch, for example. He has constructed a paddock system using the latest in fencing technology, but he did this in order to imitate nature's grazing pattern. And while he no longer has a good use for his fancy tractor, he does view the Internet as a vital tool for connecting to his extended farming community. Similarly, Jeff at Scribe Winery judiciously uses pheromone traps and sulfur in order to maintain a balance between pests and beneficial insects in his vineyard. In almost every instance where these ecologist farmers rely on technology, it is not to disrupt nature's design but rather to preserve or restore what nature has given them.

When it comes to our health, the best interventions are also those that support our body's natural patterns. Borrowing from the term "integrated pest management," it seems fair to call this new health strategy "integrated patient management." We see examples of this in every chapter of this book. Erika von Mutius, the asthma and allergy researcher, is interested in innovative immunotherapies but only in so far as they support (and do not override) the immune systems of her young patients. Justin Sonnenburg, the immunologist, and Robert Gatenby, the cancer researcher, are working on therapies that maintain a healthy ecological balance, whether between microbe and gut or between tumor cells and surrounding tissue. And Bruce McEwen, the stress expert, is interested in how to preserve a healthy stress response. To this end, he is open to using pharmaceuticals in concert with lifestyle changes and a supportive environment.

My third revelation concerns what farmers consider to be the "vital signs" of a healthy or resilient farm ecology. Three interrelated "vital signs" have reappeared time and again: diversity, synergy, and redundancy.

Diversity (or variability) can be seen in the populations of microbes in Jubilee's healthy soil, the insect populations at Scribe, and the daily activities of a pastured hen at Heartland Egg. Synergy—the whole being greater than the sum of the parts—is a theme in every chapter since each farm's success cannot be predicted simply by looking at its discrete components. Finally, redundancy, or self-sameness, describes the emergence of specific designs within each organism and throughout an entire ecosystem. Recurring patterns are a sign of a system's resilience since, in the event of a failure, one part can provide backup, or become a substitute for another. Examples include the similar loops that take place in Jubilee's soil, within the cow, and in humans, or the near identical structure of aromatic molecules synthesized by plants and by our own bodies.

These three vital signs—diversity, synergy, and redundancy—are rarely discussed within medicine where research and treatments are based on predictability, linearity, and isolation. And yet these qualities are as much a sign of vitality within our bodies as they are within a farm. Consider the biodiversity found in the intestines of healthy children eating a farm-fresh, unprocessed diet in Burkina Faso (or Bavaria), the synergy that occurs between microbes and our own immune system, or the reassuring patterns that gastroenterologist and cancer researcher Brian Reid recognized within supposedly abnormal Barrett's cells. Variability, synergy, and redundancy are also signs of health when evaluating heart rhythms, sleep rhythms, and other aspects of our physiology.[1] These concepts were not discussed in my medical training, nor had I given them much thought in my medical practice. But now that has changed. By learning from eco-farmers, I have new ways to assess and describe a healthy system.

My fourth revelation concerns the scientists and health professionals who contributed their knowledge to this book. When I started this project, I was looking to farmers to teach me about weblike models for healing. Along the way, I discovered that my own profession has its share of people who think about complexity. None of them regard themselves as "medical ecologists," but I believe they deserve this descriptor. They represent a new frontier of science, one that takes a broader, more integrated approach to health and healing. As their ranks (and funding) grow, I am certain we shall see a shift both in our understanding of health and wellness and in the structure of our health care system. And this brings me to my final insight:

By thinking like farmer ecologists, we can all make profound and lasting changes in our own health and in our lives.

Let me suggest how you might start.

DURING MY visit to Lane's Landing, I was struck by Wendell Berry's comment that the words "organic" and "sustainable" are simply shorthand for a long conversation in which all parties are mutually supportive. Now, at the end of this long journey, I understand exactly what he meant: Each farm that I visited represents a series of "conversations"—microbes talking with soil, soil talking with plants, plants talking with cows and farmers, farmers talking with the farming community. I laid out these conversations in a kind of "map" that captures some of the physical and geographic features of each farm, but mainly focuses on the messy but exciting collection of mutually beneficial activities.

As a next step, you might consider undertaking this exercise yourself—but rather than representing the dynamics of a sustainable farm, yours could be a personal health map showing all the "conversations" that contribute to your wellness. Sometimes patients hesitate when I suggest this activity. "How can I possibly do this?" they ask. "Only health experts know how my body works, and only experts can devise a plan to keep me well." In response, I reassure them that, with this ecological approach, they are the biggest expert of all.

Start externally, thinking of your relationships—not only with individuals (family, friends, lovers, colleagues, mentors, health care providers, therapists, teachers, local farmers, pets, and so on) but also with groups you are involved with (spiritual or religious, volunteer causes), your physical environment (nature, home, workplace, market, refuge), your activities and pursuits (art, cooking, books, movies, music, dance, exercise), whatever you put in your body (food, drink, medications, herbs, supplements), and other sensory experiences.

After you have drawn your external webs of wellness, move inward. Here, rather than using the standard biomedical approach of dividing the body into systems (nervous, reproductive, and urinary), you might want to think about body functions: sleep, mood and memory, digestion, sex, and so on. Next, you may want to look at the conversations you have between your internal world and the external one, for instance, between sleep quality and alcohol consumption, or between digestion and work stress.

Once you have sketched out your ecological health map, highlight those areas and relationships that you believe are thriving. Next, highlight conversations that you feel need some rebalancing—or maybe even a total overhaul. Recognizing areas that are having a negative effect on your health is often the most difficult part of starting to make things right. See if you can tackle some trouble spots on your own, or with the help of friends and loved ones. Finally, after you have done all this, please consult with a health care provider who thinks like a medical ecologist.

This last suggestion may seem odd given that to date there is no universally accepted certificate or professional degree for this brand of health professional. In reality, nurses and doctors who think like eco-farmers are more numerous than you may imagine. Perhaps by bringing your conversation map with you to an appointment, it will help to identify them. If your health care provider uses it to help you rebalance off-kilter relationships (internally and externally), then you have found the right collaborator. But if the health care provider gives you an annoyed look, pushes your map aside, and tells you to pick one problem, maybe you need to look further afield. In fact, at the risk of sounding self-serving, you might want to give that person a copy of this book.

As more and more patients demand this approach more members of my profession will begin to think like medical ecologists. And as this shift occurs we will begin to see our now unsustainable health care system become one that nurtures all of us. If still doubtful, take a look at agriculture. By shopping at farmers' markets and farm stands, by buying from CSAs, by cooking our own food, and by voting with our forks and our ballots, we have collectively managed to spawn a growing network of healthy farms and a new generation of farmers who think like farmer ecologists. It is time that we bring this same purpose and imagination to medicine.

Acknowledgments

AS I file away two years of notes and interviews, I am reminded of how many people generously contributed their time and expertise to this project. To all of them I am so grateful.

First and foremost, I owe endless thanks to the host farmers and the many scientists who undertook the task of educating me. At times they seemed to understand better than I the value of this project.

I would like to thank everyone who patiently listened and offered suggestions as I began to organize the ideas for this book, including my husband and treasured collaborator, Ross Levy; my parents, Susan and David Miller; my brother, Sam Miller; and dozens of friends and colleagues: Eleanor Jackson, Avril Swan, Michelina Labat, Grant Colfax, Rodman Rogers, Dashka Slater, Mary Ellen Hannibal, Emily Gottreich, Jared Blumenfeld, Alex Nichols, Maggie Farley, Sharon Meers, Frances Sellers, Raj Patel, and David Wallinga. Andrew Weil, Victoria Maizes, Jim Gordon, and Kathie Swift deserve a special thanks as they generously allowed me to present some of these ideas as a work-in-progress at their annual nutrition conferences.

I thank my wonderful editor, Cassie Jones, for her spot-on feedback and unflagging support, and I thank everyone else who read through portions of this manuscript: my mother, Susan Miller, who still teaches me how to write; Gary Kamiya; Michael Pollan; Deirdre English; Grace Gershuny; and all the medical researchers who reviewed their respective chapters for scientific accuracy. Amy Kristofferson and Tara Garnett, two recent UC Berkeley graduates who helped me with research and fact-checking, deserve special recognition. They are both interested in a career in medicine, and any medical school would be lucky to have them.

And finally, there are the many people who contributed to this book in indirect but equally valuable ways. My children, Arlen and Emet Levy, who forgave the late dinners and moments of distraction and offered me their own special brand of encouragement; Francesca Vietor, who gave me the keys to an idyllic writer's retreat in Bolinas; Cheryl Eccles, who regularly rescued me from my computer to run me through the Berkeley Hills; Paul Burke, who treated many of my writer's cramps; and Paul Arenstam, whose great cooking frequently offered a guilt-free substitute for my own.

To these people, and many more, I give my deepest thanks.

Notes

CHAPTER 1: JUBILEE

1. The dietary recommendations made by Allie's gastroenterologist were based on the FODMAP diet, which proposes that eliminating rapidly fermentable short-chain carbohydrates from the diet can cut down on the symptoms of gas and bloating. While this may be true, many of the foods on the FODMAP list are the very same foods that feed beneficial bacteria in the intestine. A better strategy might be to be aware of these foods and avoid eating too many of them at once, but not to avoid them altogether. For a printable FODMAP diet sheet, visit www.IBSgroup.org.
2. While some of the tests in Allie's folder were appropriate, many were not (to use medical lingo) "actionable," meaning that they had no proven clinical benefit. For some of her labs, there's still controversy about what type of sample gives the most accurate measurement: serum, plasma, stool, hair, or urine. For other tests in her portfolio, it is still not clear what constitutes an abnormal result. Take the case of vitamin D, a commonly measured substance. Although the vitamin D levels

associated with severe deficiency are well established, less is known about whether a more subtle deficiency is associated with chronic health issues, such as depression, multiple sclerosis, and certain cancers. Finally, with vitamin D, as well as many other items in Allie's folder, it's still not known whether the biomarker itself is responsible for health problems or whether it's simply a beacon for something more systemic and harder to quantify.

One might ask why tests have such an appeal—even ones with questionable utility. For health providers, a test promises something definitive to tell a suffering patient; for patients, it offers what is often a long-sought explanation for bothersome symptoms; for nutraceutical and pharmaceutical companies, tests provide opportunities to sell new products; and for hospitals (and doctors), there's revenue to be gained not only from the test itself but from a surgery or procedure that may follow. It's easy to see that the medical industry, like the farming industry, has a big stake in creating an enthusiastic market for all this sampling!

3. Most minerals used in farming and for human consumption are chelated with synthetic agents, and although some manufacturers may claim that their chelation process "perfectly imitates nature," this claim cannot be substantiated.

4. Erick, like many eco-farmers, uses a no-till approach, avoiding any equipment that churns up the earth and disrupts the life below.

5. Although several reports have shown that plants grown in biodynamic systems are more nutritious than ones raised conventionally, more research is needed to understand the effects that a biodynamic system might have on the quality of its plants as well as the health of humans who are connected to that system.

CHAPTER 2: ROCKIN' H

1. A squeeze chute, as the name implies, is a metal contraption that traps the cows so that they may receive injections or other handling.

2. The acronym for Erika von Mutius's collaborative is GABRIELA, based on its German title. The English translation is Multidisciplinary Study to Identify the Genetic and Environmental causes of Asthma in the European Community.

3. Von Mutius mentions the relatively high proportion of omega-3 fats in raw milk as another likely factor in its anti-allergen, anti-inflammatory, and immunity-boosting effect. At this point, plenty of research links consumption of these long-chain fats with a better immune response, less asthma and eczema, and fewer autoimmune problems, but what is special about raw milk is that it contains more omega-3 fats than most processed milk. This is partly because the cows themselves are likely to have eaten grass-forage rich in omega-3 rather than grain and silage, and partly because pasteurization and homogenization can destroy these protective fatty acids and limit their health-giving potential.

4. Jennifer Reeve and her colleagues at Utah State University have shown that fumigating or chemically treating a field can have long-term deleterious effects on the soil's microbial count and on biotic diversity. Also, a recent article by Kevin Forsberg in *Science* offers evidence that soil bacteria that has been exposed to manure from cows treated with antibiotics can exchange antibiotic resistant genes with common human pathogens, such as Pseudomonas. The consequence of this DNA swap is that we are seeing more-aggressive, less-treatable bacterial infections in humans.

5. A preponderance of evidence suggests that breast-feeding offers many physical and emotional advantages over bottle-feeding and that babies who are exclusively—or almost exclusively—breast-fed for the first three to six months of life have fewer food allergies and less asthma and are more likely to be normal weight as older children and as adults. Also, it is important to note that while stable dust and raw milk were statistically linked to the low rates of asthma in the Bavarian farm studies, most of those farm children are nursed for at least the first six months of life.

CHAPTER 3: HEARTLAND EGG AND ARKANSAS EGG

1. It should be noted that many of the physiologic measures used to study stress in poultry do a poor job of differentiating between the acute and the chronic. This explains why the studies I had read showed little difference in the blood levels of stress markers in indoor and outdoor birds. In chickens, behavioral measures may offer a better indicator of chronic stress. One such measure is the tonic immobility test, in which hens are caught and restrained and then timed to see how long it takes until they stop fighting. This test gives a read on a bird's degree of learned helplessness, or how helpless it feels as a result of continually being exposed to negative stimuli or stressors. Not surprisingly, hens in cages usually give in much faster than free-range hens. I thought back to how little reaction I elicited by shaking my bootied foot at the pecking hens in the Arkansas henhouse and wondered if their nervous systems were so taxed that they could not even muster a good squawk and flight.

2. Certain alleles or genetic variations can predispose an individual to anxiety and depression, while other DNA codings might make them more resilient.

CHAPTER 4: SCRIBE WINERY

1. Ironically, as you can see in the epigraphs to this chapter, it was President Nixon, the same president who called for a full-on chemical war on cancer, who directed federal agencies to embrace the holistic concept of integrated pest management.

2. Phylloxera (which originated in North America and is reported to have traveled to Europe in the specimen bag of some unsuspecting Victorian botanist) killed its victims by burrowing deep within the roots, leaving the flesh vulnerable to devastating fungal infections. Sadly, the Mariani grape growers of Vis suffered the same financial ruin as monocrop farmers everywhere whose harvest falls prey to an aggressive pest.

3. FRAC offers an international rating system for fungicides and bacteriocides. Each chemical is given a different number, corresponding to the metabolic site where it targets its prey. For example, FRAC 3 represents a group of antifungals that attack the formation of C14 sterols in the host cell. Before meeting Jeff Wheeler, I had already spent a considerable amount of time studying the FRAC chart on the Internet. The chart contains over forty-five FRAC codes, and in the far right-hand column for each code, under the heading "Comments," there is a list of all the pests and weeds that have managed to outwit that specific chemical type. Jeff's argument was that this list is constantly expanding and therefore fighting pests with synthetic chemicals is a losing proposition.
4. Recent research shows that statins (the drugs used to lower cholesterol) might also protect against esophageal cancer. As with NSAIDs, this protection is most likely attributable to a statin's anti-inflammatory effect.

CHAPTER 5: LA FAMILIA VERDE URBAN FARMS

1. As part of the White House's effort to eliminate food deserts and combat childhood obesity, it also formed an alliance with Walmart, Walgreens, and SuperValu, along with three regional chains, to open fifteen hundred new stores in low-income areas across the country.
2. La Finca del Sur on Grand Concourse is not part of Karen's neighborhood farming network—La Familia Verde—but it is the latest addition to the Bronx's extensive urban-farming coalition and whenever possible Karen goes down and volunteers. As with so many urban agricultural projects, Karen and her cofarmers had to deal with "piles of red tape" before they could plant their first seed. First, they had to obtain the rights to the property. Karen told me that each quadrant of La Finca had a different owner: the Departments of Parks, the Department of Transportation, Metro-North Railroad, and a private family. Just getting all of the parties to agree that the land could be used for community agriculture turned out to be a Herculean task. Next, they had to make

sure that the soil was safe for growing food. Amazingly, despite being bounded by the congested Grand Concourse, the interstate, and the Metro-North commuter train tracks, the levels for heavy metals and other toxins were surprisingly low—in fact, no worse than your average farm soil. Karen attributed this to the thick cover of leaf-fall and grass that have continuously blanketed the area.

3. A limitation of this study was that it did not indicate whether giving up gardening *caused* the decline or whether it was an early indicator of disintegrating health.

CHAPTER 6: MORNING MYST

1. I can't help but wonder how much all of those synthetic scents in our environment—such as Chanel No. 5, Febreze air fresheners, and Bounce—have overwhelmed our olfactory systems and distorted our appreciation for natural, subtle beauty.

PREFACE: SETTING OUT

Berry, Wendell. "The Body and the Earth." In *The Unsettling of America: Culture and Agriculture*, pp. 97–140. San Francisco: Sierra Club Books, 1977.

INTRODUCTION: KICKING OVER THE TRACES

The statistics on organic and sustainable farms are taken from the Organic Farming Research Foundation (http://ofrf.org/organic-faqs) and U.S. Department of Agriculture (USDA.gov) websites.

Fukuoka, Masanobu. *The One-Straw Revolution: An Introduction to Natural Farming*. New York: NYRB Classics, 2009.

Gershuny, Grace, and Joe Smillie. *The Soul of Soil: A Soil-Building Guide for Master Gardeners and Farmers*. White River Junction, VT: Chelsea Green Publishing, 1999.

Giovannucci, Daniele, Sara Scherr, Danielle Nierenberg, Charlotte Hebebrand, Julie Shapiro, Jeffrey Milder, and Keith Wheeler. 2012. *Food and Agriculture: The Future of Sustainability*. A strategic input to the

Sustainable Development in the 21st Century (SD21) Project. New York: United Nations Department of Economic and Social Affairs, Division for Sustainable Development.

Howard, Albert. *The Soil and Health: A Study of Organic Agriculture*. Lexington: University Press of Kentucky, 2007.

King, F. H. *Farmers of Forty Centuries: Organic Farming in China, Korea, and Japan*. Mineola, NY: Dover Publications, 2004 (originally published 1911).

McKibben, Bill. *Eaarth: Making a Life on a Tough New Planet*. New York: Henry Holt and Co., 2010.

Pollan, Michael. *The Omnivore's Dilemma: A Natural History of Four Meals*. New York: Penguin Press, 2006.

Shiva, Vandana. *Stolen Harvest: The Hijacking of the Global Food Supply*. Cambridge, MA: South End Press, 2000.

Medicine and Farming: A Common History

Keegan, John. *The First World War*. New York: Knopf Doubleday, 1999.

Lindberg, David C. *The Beginnings of Western Science*. Chicago: University of Chicago Press, 1992.

U.S. Department of the Army. Office of the Surgeon General. "Medical Aspects of Chemical and Biological Warfare" (1997). In part 1 of *Textbook of Military Medicine*. Available at: http://www.bvsde.paho.org/cursode/fulltext/chebio.pdf. (accessed October 25, 2012).

Complexity Thinking in Medicine

Capra, Fritjof. *The Hidden Connections: Integrating the Biological, Cognitive, and Social Dimensions of Life into a Science of Sustainability*. New York: Doubleday, 2002.

Jayasinghe, S. "Conceptualizing Population Health: From Mechanistic Thinking to Complexity Science." *Emerging Themes in Epidemiology* 8(1, 2011): 2.

Maizes, V., D. Rakel, and C. Niemiec. "Integrative Medicine and Patient-Centered Care." *Explore* 5(2009): 277–89.

Plsek, P. E., and T. Greenhalgh. "Complexity Science: The Challenge of Complexity in Health Care." *British Medical Journal* 323(7313, 2001): 625–28.

Sturmberg, J. P. "Systems and Complexity Thinking in General Practice." *Australian Family Physician* 36(3, 2007): 170–73.

Weil, Andrew. *Why Our Health Matters: A Vision of Medicine That Can Transform Our Future*. New York: Hudson Street Press, 2009.

West, G. B. "The Importance of Quantitative Systemic Thinking in Medicine." *Lancet* 379(2012): 1551–59.

West, G. B., and A. Bergman. "Toward a Systems Biology Framework for Understanding Aging and Health Span." *Journal of Gerontology* 64A(2, 2009): 205–8.

Yawar, A. "Can Doctors Think?" *Lancet* 372(2008): 1285–86.

Examples of How Medicine Is Slow to Change

Niino, Y. "The Increasing Cesarean Rate Globally and What We Can Do About It." *BioScience Trends* 5(2011): 139–50.

Timor-Tritsch, I. E., and A. Monteagudo. "Unforeseen Consequences of the Increasing Rate of Cesarean Deliveries: Early Placenta Accreta and Cesarean Scar Pregnancy: A Review." *American Journal of Obstetrics and Gynecology* 207(2012): 14–29.

Vickers, A. J., M. J. Roobol, and H. Lilja. "Screening for Prostate Cancer: Early Detection or Overdetection?" *Annual Review of Medicine* 63(2012): 161–70.

CHAPTER 1: JUBILEE

The Golden Mean of Soil

Albrecht, William A. *The Albrecht Papers*, vol. 1, *Foundation Concepts*. Austin, TX: Acres USA, 2011.

Reid, I. R., and M. J. Bolland. "Calcium Supplements: Bad for the Heart?" *Heart* 98(12, 2012): 895–96.

Testing Reconsidered

Bilinksi, K., and S. Boyages. "The Rise and Rise of Vitamin D Testing." *British Medical Journal* 345(2012): e4743–44.

Carter, G. D. "25-Hydroxyvitamin D Assays: The Quest for Accuracy." *Clinical Chemistry* 55(7, 2009): 1300–1302.

Elston, D. M. "Commentary: Iron Deficiency and Hair Loss: Problems with Measurement of Iron." *Journal of the American Academy of Dermatology* 63(2010): 1077–82.

Glendenning, P., and C. Inderjeeth. "Screening for Vitamin D Deficiency: Defining Vitamin D Deficiency, Target Thresholds of Treatment, and Estimating the Benefits of Treatment." *Pathology* 44(2, 2012): 160–65.

Ioannidis, J.P.A. "Limits to Forecasting in Personalized Medicine: An Overview." *International Journal of Forecasting* 25(2009): 773–83.

Joseph, J., D. E. Handy, and J. Loscalzo. "Quo Vadis: Whither Homocysteine Research?" *Cardiovascular Toxicology* 9(2009): 53–63.

Lind, B. K., et al. "Comparison of Health Care Expenditures Among Insured Users and Nonusers of Complementary and Alternative Medicine in Washington State: A Cost Minimization Analysis." *Journal of Alternative and Complementary Medicine* 16(2010): 411–17.

McShane, L. M. "Statistical Challenges in the Development and Evaluation of Marker-Based Clinical Tests." *BMC Medicine* 19(2012): 52.

Morrison, Jessica. "Vitamin D Gets Frequent Testing, but the Results Are a Bit Quizzical." *Chicago Tribune*, July 18, 2012.

Rabin, Roni Caryn. "Doctor Panels Recommend Fewer Tests for Patients." *New York Times*, April 4, 2012.

Welsh, P., C. J. Packard, and N. Sattar. "Novel Antecedent Plasma Biomarkers of Cardiovascular Disease: Improved Evaluation Methods and Comparator Benchmarks Raise the Bar." *Current Opinion in Lipidology* 19(6, 2008): 563–71.

Wong, W. B., et al. "Cost Effectiveness of Pharmacogenomics: A Critical and Systematic Review." *Pharmacoeconomics* 28(2010): 1001–13.

Cone Spreaders and Bags of Vitamins

Burton, G. W., et al. "Human Plasma and Tissue Alpha-Tocopherol Concentrations in Response to Supplementation with Deuterated Natural and Synthetic Vitamin E." *American Journal of Clinical Nutrition* 67(1998): 669–84.

Eilat-Adar, S., and U. Goldbourt. "Nutritional Recommendations for Preventing Coronary Heart Disease in Women: Evidence Concerning Whole Foods and Supplements." *Nutrition, Metabolism, and Cardiovascular Diseases* 20(6, 2010): 459–66.

Fairfield, K. M. "Vitamins for Chronic Disease Prevention in Adults: Scientific Review." *Journal of the American Medical Association* 287(2002): 3116–26.

Kelly, P., et al. "Unmetabolized Folic Acid in Serum: Acute Studies in Subjects Consuming Fortified Food and Supplements." *American Journal of Clinical Nutrition* 65(1997): 1790–95.

Lockwood, G. B. "The Quality of Commercially Available Nutraceutical Supplements and Food Sources." *Journal of Pharmacy and Pharmacology* 63(2011): 3–10.

Miller, E. R., III, et al. "Review Meta-Analysis: High-Dosage Vitamin E Supplementation May Increase Mortality." *Annals of Internal Medicine* 142(1, 2005): 37–46.

Minich, D. M., and J. S. Bland. "Dietary Management of the Metabolic Syndrome Beyond Macronutrients." *Nutrition Reviews* 66(8, 2008): 429–44.

Mulholland, C., and D. J. Benford. "What Is Known About the Safety of Multivitamin-Multimineral Supplements for the Generally Healthy Population? Theoretical Basis for Harm." *American Journal of Clinical Nutrition* 85(2007): 318S–22S.

An Evening with Rudolf Steiner

Steiner, Rudolf. *Agriculture Course: The Birth of the Biodynamic Method.* Herndon, VA: Steiner Books, 2004.

———. *What Is Biodynamics? A Way to Heal and Revitalize the Earth.* Herndon, VA: Steiner Books, 2005.

Wright, Hilary. *Biodynamic Gardening for Health and Taste*. Edinburgh, U.K.: Floris Books, 2009.

The Law of the Return, or the Eternal Dance of the Microbes

Badri, D. V., et al. "Rhizosphere Chemical Dialogues: Plant-Microbe Interactions." *Current Opinion in Biotechnology* 20(2009): 642–50.

Center for Food Safety. "Genetically Engineered Crops, 2012." Available at: http://www.centerforfoodsafety.org/campaign/genetically-engineered-food/crops/ (accessed September 11, 2012).

Dirt! The Movie. Directed by Bill Benenson, Gene Rosow, and Eleanore Dailly. New Video Studios, 2009 (DVD).

Elless, M. P., et al. "Plants as a Natural Source of Concentrated Mineral Nutritional Supplements." *Food Chemistry* 71(2, 2000): 181–88.

Graham, R. D., and J.C.R. Stangoulis. "Comparative Trace Element Nutrition, Trace Element Uptake and Distribution in Plants." *Journal of Nutrition* 133(2003): 1502–05.

Grigg, Gary. "Fertilization with Foliar Absorbent Nutrients." Grigg Brothers News, available at: http://www.griggbros.com/index.php/research/techbulletins/fertilization-with-foliar-absorbed-nutrients (accessed September 12, 2012).

Haertl, E. J. "Metal Chelates in Plant Nutrition." *Agricultural and Food Chemistry* 11(2, 1963): 108–11.

Reeve, J. R., et al. "Effects of Soil Type and Farm Management on Soil Ecological Functional Genes and Microbial Activities." *ISME Journal* 4(2010): 1099–1107.

Sillanpää, M., and K. Pirkanniemi. "Recent Developments in Chelate Degradation." *Environmental Technology* 22(7, 2001): 791–801.

Symphony of the Soil. Directed by Deborah Koons. Lily Films, 2012 (DVD).

Teixeira, L.C.R.S., et al. "Bacterial Diversity in Rhizosphere Soil from Antarctic Vascular Plants of Admiralty Bay, Maritime Antarctica." *ISME Journal* 4(2010): 989–1001.

U.S. Department of Agriculture. "Amino Acids." In NOSB Materials Database, October 12, 2007. Available at: http://www.ams.usda.gov/AMSv1.0/getfile?dDocName=STELPRDC5066962 (accessed September 12, 2012).

Eco-Farming vs. Organic vs. Conventional: How Do They Compare?

Asami, D. K., et al. "Comparison of the Total Phenolic and Ascorbic Acid Content of Freeze-Dried and Air-Dried Marionberry, Strawberry, and Corn Grown Using Conventional, Organic, and Sustainable Agricultural Practices." *Journal of Agricultural and Food Chemistry* 51(2003): 1237–41.

Barrett, D. M., et al. "Qualitative and Nutritional Differences in Processing Tomatoes Grown Under Commercial Organic and Conventional Production Systems." *Journal of Food Science* 72(2007): C441–51.

Bavec, M., et al. "Influence of Industrial and Alternative Farming Systems on Contents of Sugars, Organic Acids, Total Phenolic Content, and the Antioxidant Activity of Red Beet (*Beta vulgaris L. ssp. vulgaris Rote Kugel*)." *Journal of Agricultural and Food Chemistry* 58(2010): 11825–31.

Bittman, Mark. "Eating Food That's Better for You, Organic or Not." *New York Times,* March 21, 2009.

Carpenter-Boggs, L., A. Kennedy, and J. Reganold. "Organic and Biodynamic Management: Effects on Soil Biology." *Soil Science Society of America Journal* 64(2000): 1651–59.

Cordero-Bueso, G., et al. "Influence of the Farming System and Vine Variety on Yeast Communities Associated with Grape Berries." *International Journal of Food Microbiology* 145(2011): 132–39.

Crinnion, W. J. "Organic Foods Contain Higher Levels of Certain Nutrients, Lower Levels of Pesticides, and May Provide Health Benefits for the Consumer." *Alternative Medicine Review: A Journal of Clinical Therapeutics* 15(2010): 4–12.

Dangour, A. D., et al. "Nutrition-Related Health Effects of Organic Foods: A Systematic Review." *American Journal of Clinical Nutrition* 92(2010): 203–10.

McCullum, C. "Using Sustainable Agriculture to Improve Human Nutrition and Health." *Journal of Community Nutrition* 6(2004): 18–25.

Reeve, J. R., Carpenter-Boggs, L., Reganold, J. P., York, A. L., and Brinton, W. F. "Influence of Biodynamic Preparations on Compost Development and Resultant Compost Extracts on Wheat Seedling Growth." *Bioresource Technology* 101(2010): 5658–66

Reganold, J. P., et al. "Performance of and Financial Soil Quality Farms in New Zealand and Conventional Biodynamic of Agriculture." *Advancement of Science* 260(2011): 344–49.

Rosenthal, Elisabeth. "Organic Agriculture May Be Outgrowing Its Ideals." *New York Times*, December 30, 2011.

Spaccini, R., Mazzei, P., Squartini, A, Giannattasio, M., and Piccolo, A. "Molecular Properties of a Fermented Manure Preparation Used as Field Spray in Biodynamic Agriculture." *Environmental Science and Pollution Research International* 19(2012): 4214–25

Van Wesemael, B., et al. "Agricultural Management Explains Historic Changes in Regional Soil Carbon Stocks." *Proceedings of the National Academy of Sciences of the United States of America* 107(2010): 14926–30.

Williams, C. M. "Nutritional Quality of Organic Food: Shades of Grey or Shades of Green?" *Proceedings of the Nutrition Society* 61(2007): 19–24.

Statistics on Food-Borne Illness

Centers for Disease Control and Prevention (CDC). "CDC Estimates of Foodborne Illness in the United States." Available at: http://www.cdc.gov/foodborneburden/2011-foodborne-estimates.html (accessed September 13, 2012).

The Gnotobiotic Farmer

Davis, C. D., and J. A. Milner. "Prevention." *Journal of Nutrition* 20(2010): 743–52.

De Filippo, C., et al. "Impact of Diet in Shaping Gut Microbiota Revealed by a Comparative Study in Children from Europe and Rural Africa." *Proceedings of the National Academy of Sciences of the United States of America* 107(2010): 14691–96.

Goodman, A. L., and J. I. Gordon. "Our Unindicted Coconspirators: Human Metabolism from a Microbial Perspective." *Cell Metabolism* 12(2010): 111–16.

Hehemann, J. H., et al. "Transfer of Carbohydrate-Active Enzymes from Marine Bacteria to Japanese Gut Microbiota." *Nature* 464 (2010): 908–12.

Hong, H., et al. "*Bacillus subtilis* Isolated from the Human Gastrointestinal Tract." *Research in Microbiology* 160(2009): 134–43.

Lee, J. H., and D. J. O'Sullivan. "Genomic Insights into Bifidobacteria." *Microbiology and Molecular Biology Reviews* 74(2010): 378–416.

Ley, R. E. "Obesity and the Human Microbiome." *Current Opinion in Gastroenterology* 26(2010): 5–11.

Ley, R. E., et al. "Evolution of Mammals and Their Gut Microbes." *Science* 320(2008): 1647–51.

Ley, R. E., et al. "Worlds Within Worlds: Evolution of the Vertebrate Gut Microbiota." *Nature Reviews Microbiology* 6(2008): 776–88.

Musso, G., R. Gambino, and M. Cassader. "Obesity, Diabetes, and Gut Microbiota: The Hygiene Hypothesis Expanded?" *Diabetes Care* 33(2010): 2277–84.

Sonnenburg, J. L. "Genetic Pot Luck." *Nature* 464(8, 2010): 837–38.

Sonnenburg, J. L., L. T. Angenent, and J. I. Gordon. "Getting a Grip on Things: How Do Communities of Bacterial Symbionts Become Established in Our Intestine?" *Nature Immunology* 5(2004): 569–73.

Turnbaugh, P. J., et al. "An Obesity-Associated Gut Microbiome with Increased Capacity for Energy Harvest." *Nature* 444(2006): 1027–31.

Xu, J., and J. I. Gordon. "Honor Thy Symbionts." *Proceedings of the National Academy of Sciences of the United States of America* 100(2003): 10452–59.

Zocco, M., M. E. Ainora, and G. Gasbarrini. "Bacteroides Thetaiotaomicron in the Gut: Molecular Aspects of Their Interaction." *Digestive and Liver Disease: Official Journal of the Italian Society of Gastroenterology and the Italian Association for the Study of the Liver* 39(2007): 707–12.

Fighting Extinction from Within: The Benefit of Prebiotics and Probiotics

Bosscher, D., et al. "Food-Based Strategies to Modulate the Composition of the Intestinal Microbiota and Their Associated Health Effects." *Pharmacology* 60(suppl. 6, 2009): 5–11.

Dotterud, C. K., et al. "Probiotics in Pregnant Women to Prevent Allergic Disease: A Randomized, Double-Blind Trial." *British Journal of Dermatology* 163(2010): 616–23.

Gibson, G. R. "Prebiotics as Gut Microflora Management Tools." *Journal of Clinical Gastroenterology* 42(suppl. 2, 2008): 75–79.

Hedin, C., K. Whelan, and J. O. Lindsay. "Evidence for the Use of Probiotics and Prebiotics in Inflammatory Bowel Disease: A Review of Clinical Trials."*Proceedings of the Nutrition Society* 66(2007): 307–15.

Kinouchi, F., et al. "A Soy-Based Product Fermented by *Enterococcus faecium* and *Lactobacillus helveticus* Inhibits the Development of Murine Breast Adenocarcinoma." *Food and Chemical Toxicology: International Journal Published for the British Industrial Biological Research Association* 50(11, 2012): 4144–48.

Lane, J., et al. "The Food Glycome: A Source of Protection Against Pathogen Colonization in the Gastrointestinal Tract." *International Journal of Food Microbiology* 142(2010): 1–13.

Lee, S., et al. "Antidiabetic Effect of *Morinda citrifolia* (Noni) Fermented by *Cheonggukjang* in KK-A(y) Diabetic Mice." *Evidence-Based Complementary and Alternative Medicine* 2012(2012): 1–8.

Lemon, K., et al. "Microbiota-Targeted Therapies: An Ecological Perspective." *Science Translational Medicine* 4(137, 2012): 1–8.

Li, S., et al. "Antioxidant Activity of *Lactobacillus plantarum* Strains Isolated from Traditional Chinese Fermented Foods." *Food Chemistry* 135(2012): 1914–19.

Morrow, L., V. Gogineni, and M. Malesker. "Probiotic, Prebiotic, and Synbiotic Use in Critically Ill Patients." *Current Opinion in Critical Care* 18(2, 2012): 186–91.

Ng, S. C., et al. "Mechanisms of Action of Probiotics: Recent Advances." *Inflammatory Bowel Diseases* 15(2009): 300–310.

Parker-Pope, Tara. "Probiotics: Looking Underneath the Yogurt Label." *New York Times*, September 29, 2009.

Rafter, J., et al. "Dietary Synbiotics Reduce Cancer Risk Factors in Polypectomized and Colon Cancer Patients." *American Journal of Clinical Nutrition* 8(25, 2007): 488–96.

Reid, G., K. Anukam, and T. Koyama. "Probiotic Products in Canada with Clinical Evidence: What Can Gastroenterologists Recommend?" *Canadian Journal of Gastroenterology* 22(2, 2008): 169–75.

Avoid Killing Off Good Bacteria

Blaser, M. "Stop the Killing of Beneficial Bacteria." *Nature* 476(2001): 393–94.

Dethlefsen, L., and D. A. Relman. "Incomplete Recovery and Individualized Responses of the Human Distal Gut Microbiota to Repeated Antibiotic Perturbation." *PNAS* 108(suppl. 1, 2011): 4554–61.

Forsberg, K. J., et al. "The Shared Antibiotic Resistome of Soil Bacteria and Human Pathogens." *Science* 337(2012): 1107–11.

Huse, S., et al. "Exploring Microbial Diversity and Taxonomy Using SSU rRNA Hypervariable Tag Sequencing." *PLoS Genetics* 4(11, 2008): 2383–2400.

Koenig, J. E., et al. "Succession of Microbial Consortia in the Developing Infant Gut Microbiome." *Proceedings of the National Academy of Sciences of the United States of America* 108(suppl. 1, 2011): 4578–85.

Looft, T., and H. K. Allen. "Collateral Effects of Antibiotics on Mammalian Gut Microbiomes." *Gut Microbes* 3(5, 2012): 1–5.

Farm Communities and Health

Butler, Kiera. "Econundrum: Which Gardening Moves Burn Most Calories?" *Mother Jones*, April 5, 2010.

Coila, Bridget. "How Many Calories Are Burned Gardening?" *Livestrong*, July 26, 2011.

Flöistrup, H., et al. "Allergic Disease and Sensitization in Steiner School Children." *Journal of Allergy and Clinical Immunology* 117(2006): 59–66.

Kotz, Deborah. "Host of Health Benefits Attributed to Sunlight." *U.S. News and World Report*, June 24, 2008.

Loyola University Health System. "Vitamin D Lifts Mood During Cold Weather Months, Researchers Say." *Science Daily*, March 8, 2010.

CHAPTER 2: ROCKIN' H

The Holistic Farmer

Brooks, C., N. Pearce, and J. Douwes. "The Hygiene Hypothesis in Allergy and Asthma: An Update." *Current Opinion in Allergy and Clinical Immunology* (October 25, 2012) (epub ahead of print).

———. "The Hygiene Hypothesis in Allergy and Asthma: An Update." *Current Opinion in Allergy and Clinical Immunology* 13(1, 2013): 70–77.

Holmes, Cody. *Ranching Full-Time on Three Hours a Day: Real-World Validation of Holistic Systems for Stockmen*. Austin, TX: Acres USA, 2011.

Lalman, D., and R. Smith. "Effects of Preconditioning on Health, Performance, and Prices of Weaned Calves." Oklahoma Cooperative Extension Service. Available at: http://oqbn.okstate.edu/resources/ANSI-3529web.pdf.

Savory, Allan. *Holistic Management: A New Framework for Decision Making*. Washington, DC: Island Press, 1998.

Voisin, Andre. *Grass Productivity*. Washington, DC: Island Press, 1988.

Risk of Food-Borne Illness in Different Food Categories

Centers for Disease Control and Prevention. "CDC Estimates of Foodborne Illness in the United States: CDC 2011 Estimates: Findings." 2012. Available at: http://www.cdc.gov/foodborneburden/2011-foodborne-estimates.html (accessed September 13, 2012).

———. "CDC Estimates of Foodborne Illness in the United States: Trends in Foodborne Illness in the United States, 1996–2010." 2012. Available at: http://www.cdc.gov/foodborneburden/trends-in-foodborne-illness.html (accessed September 13, 2012).

Food and Drug Administration, Center for Food Safety and Applied Nutrition, Food Safety and Inspection Service. "Quantitative Assessment of Relative Risk to Public Health from Foodborne *Listeria monocytogenes* Among Selected Categories of Ready-to-Eat Foods." Interpretive Summary, Table 4. Washington, DC: U.S. Government Printing Office, 2003.

The Sensible Professor von Mutius

Ege, M. J., et al. "Exposure to Environmental Microorganisms and Childhood Asthma." *New England Journal of Medicine* 364(8, 2011): 701–9.

Ege, M. J., et al. "Prenatal Farm Exposure Is Related to the Expression of Receptors of the Innate Immunity and to Atopic Sensitization in School-Age Children." *Journal of Allergy and Clinical Immunology* 117(4, 2006): 817–23.

Farm Milk

Braun-Fahrländer, C., and E. von Mutius. "Can Farm Milk Consumption Prevent Allergic Diseases?" *Clinical and Experimental Allergy: Journal of the British Society for Allergy and Clinical Immunology* 41(1, 2011): 29–35.

Dupont, D., et al. "Food Processing Increases Casein Resistance to Simulated Infant Digestion." *Molecular Nutrition and Food Research* 54(11, 2010): 1677–89.

Haug, A., A. T. Høstmark, and O. M. Harstad. "Bovine Milk in Human Nutrition: A Review." *Lipids in Health and Disease* 6(2007): 25.

Marcobal, A., et al. "Bacteroides in the Infant Gut Consume Milk Oligosaccharides via Mucus-Utilization Pathways." *Cell Host and Microbe* 10(5, 2011): 507–14.

Sela, D., and D. Mills. "Nursing Our Microbiota: Molecular Linkages Between Bifidobacteria and Milk Oligosaccharides." *Trends in Microbiology* 18(7, 2010): 298–307.

Waser, M., et al. "Inverse Association of Farm Milk Consumption with Asthma and Allergy in Rural and Suburban Populations Across Europe." *Clinical and Experimental Allergy: Journal of the British Society for Allergy and Clinical Immunology* 37(5, 2007): 661–70.

Garbage In, Garbage Out

Derbyshire, David. "It's Not All White: The Cocktail of Up to 20 Chemicals in a Glass of Milk." *Daily Mail,* July 7, 2011.

Food and Agriculture Organization of the United Nations. "Milk Hygiene." Available at: http://www.fao.org/docrep/004/T0218E/T0218E03.htm (accessed September 28, 2012).

Haskell, M. J., et al. "The Effect of Organic Status and Management Practices on Somatic Cell Counts on UK Dairy Farms." *Journal of Dairy Science* 92(8, 2009): 3775–80.

Bavarian Stables of Plenty

Debarry J., et al. "*Acinetobacter lwoffii* and *Lactococcus lactis* Strains Isolated from Farm Cowsheds Possess Strong Allergy-Protective Properties." *Journal of Allergy and Clinical Immunology* 119(6, 2007): 1514–21.

Ege, M. J., et al. "Not All Farming Environments Protect Against the Development of Asthma and Wheeze in Children." *Journal of Allergy and Clinical Immunology* 119(5, 2007): 1140–47.

Forsberg, K. J., et al. "The Shared Antibiotic Resistome of Soil Bacteria and Human Pathogens." *Science* 337(2012): 1107–11.

Leynaert, B., et al. "Does Living on a Farm During Childhood Protect Against Asthma, Allergic Rhinitis, and Atopy in Adulthood?" *Critical Care Medicine* 164(2001): 1829–34.

Peters, M., et al. "Inhalation of Stable Dust Extract Prevents Allergen-Induced Airway Inflammation and Hyperresponsiveness." *Thorax* 61(2, 2006): 134–39.

Riedler, J., et al. "Austrian Children Living on a Farm Have Less Hay Fever, Asthma, and Allergic Sensitization." *Clinical and Experimental Allergy: Journal of the British Society for Allergy and Clinical Immunology* 30(2, 2000): 194–200.

Riedler, J., et al. "Exposure to Farming in Early Life and Development of Asthma and Allergy: A Cross-sectional Survey." *Lancet* 358(2001): 1129–33.

Smits, H., et al. "Cholera Toxin B Suppresses Allergic Inflammation Through Induction of Secretory IgA." *Mucosal Immunology* 2(4, 2009): 331–39.

Worms Too

Am, C., P. Bager, and S. Kumar. "Helminth Therapy (Worms) for Allergic Rhinitis" (review). *The Cochrane Collaboration* 4(2012): 1–46.

Lynch, N. R., et al. "Effect of Anthelmintic Treatment on the Allergic Reactivity of Children in a Tropical Slum." *Journal of Allergy and Clinical Immunology* 92(3, 1993): 404–11.

Okada, H., et al. "The 'Hygiene Hypothesis' for Autoimmune and Allergic Diseases: An Update." *Clinical and Experimental Immunology* 160(1, 2010): 1–9.

Whelan, R. K., S. Hartmann, and S. Rausch. "Nematode Modulation of Inflammatory Bowel Disease." *Protoplasma* 249(4, 2011): 871–86.

Wills-Karp, M., J. Santeliz, and C. L. Karp. "The Germless Theory of Allergic Disease: Revisiting the Hygiene Hypothesis." *Nature Reviews Immunology* 1(2001): 69–75.

Weaning Off Weaning

Lalman, D., and R. Smith. "Effects of Preconditioning on Health, Performance, and Prices of Weaned Calves." Oklahoma Cooperative Extension Service. Available at: http://oqbn.okstate.edu/resources/ANSI-3529web.pdf.

University of Pennsylvania, College of Agricultural Sciences. "Early Weaning Strategies." Available at: http://extension.psu.edu/animals/dairy/health/nutrition/calves/feeding/early-weaning-strategies (accessed September 28, 2012).

Nursing Our Palates

Beauchamp, G. K., and J. Mennella. "Flavor Perception in Human Infants: Development and Functional Significance." *Digestion* 83(suppl. 1, 2011): 1–6.

———. "Early Flavor Learning and Its Impact on Later Feeding Behavior." *Journal of Pediatric Gastroenterology and Nutrition* 48(10, 2009): S25–30.

Birch L. L., et al. "Clean Up Your Plate: Effects of Child Feeding Practices on the Condition of Meal Size." *Learning and Motivation* 18(1987): 301–17.

Ellwood, P., et al. "Diet and Asthma, Allergic Rhinoconjunctivitis, and Atopic Eczema Symptom Prevalence: An Ecological Analysis of the International Study of Asthma and Allergies in Childhood (ISAAC) Data: ISAAC Phase One Study Group." *European Respiratory Journal: Official Journal of the European Society for Clinical Respiratory Physiology* 17(3, 2001): 436–43.

Maynard M., et al. "Fruit, Vegetables, and Antioxidants in Childhood and Risk of Adult Cancer: The Boyd Orr Cohort." *Journal of Epidemiology and Community Health* 57(3, 2003): 218–25.

Mennella, J., C. P. Jagnow, and G. K. Beauchamp. "Prenatal and Postnatal Flavor Learning by Human Infants." *Pediatrics* 107(6, 2001): 1–8.

———. "Vegetable Acceptance by Infants: Effects of Formula Flavors." *Early Human Development* 82(7, 2006): 463–68.

Mennella, J., et al. "The Timing and Duration of a Sensitive Period in Human Flavor Learning: A Randomized Trial." *American Journal of Clinical Nutrition* 93(2011): 1019–24.

Ness, A. R., et al. "Diet in Childhood and Adult Cardiovascular and All Cause Mortality: The Boyd Orr Cohort." *Heart* 91(7, 2005): 894–98.

Rohlfs, P. "Flavor Exposure During Sensitive Periods of Development as a Key Mechanism of Flavor Learning: Implications for Future Research." *American Journal of Clinical Nutrition* 93(2011): 909–10.

Schwartz, C., et al. "Development of Healthy Eating Habits Early in Life: Review of Recent Evidence and Selected Guidelines." *Appetite* 57(3, 2011): 796–807.

Schwartz, C., et al. "The Role of Taste in Food Acceptance at the Beginning of Complementary Feeding." *Physiology and Behavior* 104(4, 2011): 646–52.

Environment and Childhood Food Choices

Bergmann K. E., R. L. Bergmann, and R. von Kries. "Early Determinants of Childhood Overweight and Adiposity in a Birth Cohort Study: Role of Breastfeeding." *International Journal of Obesity and Related Metabolic Disorders* 27(2, 2003): 162–72.

Erinosho, T. O., et al. "Caregiver Food Behaviours Are Associated with Dietary Intakes of Children Outside the Child-Care Setting." *Public Health Nutrition* 10(2012): 1–10.

Gillman M. W., S. L. Rifas-Shiman, and C.A.J. Camargo. "Risk of Overweight Among Adolescents Who Were Breastfed as Infants." *Journal of the American Medical Association* 285(19, 2011): 2461–67.

Harris, D., et al. "Farm to Institution: Creating Access to Healthy Local and Regional Foods." *Advances in Nutrition* 2(2012): 343–49.

Rauzon, S., et al. "An Evaluation of the School Lunch." Berkeley: University of California at Berkeley, 2010. Available at: http://edibleschoolyard.org/sites/default/files/file/An_Evaluation_of_the_School_Lunch_Initiative_Final%20Report_9_22_10.pdf (accessed October 25, 2012).

Wyse, R., et al. "Associations Between Characteristics of the Home Food Environment and Fruit and Vegetable Intake in Preschool Children: A Cross-sectional Study." *BMC Public Health* 11(2011): 938.

Allergy-Proofing Frankie

American Academy of Pediatrics. "Breastfeeding FAQs." Available at: http://www2.aap.org/breastfeeding/faqsbreastfeeding.html (accessed September 27, 2012).

American Academy of Pediatrics and American Academy of Family Physicians. "Diagnosis and Management of Acute Otitis Media." *Pediatrics* 113(4, 2004): 1451–65.

Bodewes, R., et al. "Annual Vaccination Against Influenza Virus Hampers Development of Virus-Specific CD8+ T Cell Immunity in Children." *Journal of Virology* 85(22, 2011): 11995–12000.

Boehm, G., and B. Stahl. "Oligosaccharides from Milk." *Journal of Nutrition* 137(3, 2007): 847S–49S.

Buonaguro, L., et al. "Systems Biology Applied to Vaccine and Immunotherapy Development." *BMC Systems Biology* 5(2011): 146.

Demicheli, V., et al. "Vaccines for Measles, Mumps, and Rubella in Children" (review). *The Cochrane Collaboration* 2(2012): 1–161.

Dotterud, C. K., et al. "Probiotics in Pregnant Women to Prevent Allergic Disease: A Randomized, Double-Blind Trial." *British Journal of Dermatology* 163(3, 2010): 616–23.

Du Toit, G., et al. "Early Consumption of Peanuts in Infancy Is Associated with a Low Prevalence of Peanut Allergy." *Journal of Allergy and Clinical Immunology* 122(5, 2008): 984–91.

Katz, Y., et al. "Early Exposure to Cow's Milk Protein Is Protective Against IgE-Mediated Cow's Milk Protein Allergy." *Journal of Allergy and Clinical Immunology* 126(1, 2010): 77–82.

Kwok, R. "The Real Issues in Vaccine Safety." *Nature* 473(2011): 436–38.

Pelucchi, C., et al. "Probiotics Supplementation During Pregnancy or Infancy for the Prevention of Atopic Dermatitis: A Meta-analysis." *Epidemiology* 23(3, 2012): 402–14.

Sarrell, E. M., H. A. Cohen, and E. Kahan. "Naturopathic Treatment for Ear Pain in Children." *Pediatrics* 111(5, 2012): e574–79.

Sausenthaler, S., et al. "Early Diet and the Risk of Allergy: What Can We Learn from the Prospective Birth Cohort Studies GINIplus and LISAplus?" *American Journal of Clinical Nutrition* 94(suppl., 2011): 2012–17.

Zaknun, D., et al. "Potential Role of Antioxidant Food Supplements, Preservatives, and Colorants in the Pathogenesis of Allergy and Asthma." *International Archives of Allergy and Immunology* 157(2, 2012): 113–24.

CHAPTER 3: HEARTLAND EGG AND ARKANSAS EGG

Stark, M. "The Sandpile Model: Optimal Stress and Hormesis." *Dose Response* 10(1, 2012): 66–74. Epub ahead of print, October 14, 2011.

Comparing Arkansas Egg and Heartland Egg

Chen, B. L., K. L. Haith, and B. Mullens. "A Beak Condition Drives Abundance and Grooming-Mediated Competitive Asymmetry in a Poultry Ectoparasite Community." *Parasitology* 138(2011): 748–57.

Dawkins, M., C. Donnelly, and T. Jones. "Chicken Welfare Is Influenced More by Housing Conditions Than by Stocking Density." *Nature* 427(6972, 2004): 342–44.

Greene, Robert. "Which Came First: Chickens, Eggs, or Proposition 2?" *Los Angeles Times*, July 8, 2010.

Jacob, J., and T. Pescatore. "Blood Spot Eggs." University of Kentucky College of Agriculture. Available at: http://www2.ca.uky.edu/afspoultry-files/pubs/Blood_spot_eggs.pdf (accessed October 11, 2012).

Lay, D. C., et al. "Hen Welfare in Different Housing Systems." *Poultry Science* 90(2011): 278–94.

Mugnai, C., et al. "The Effects of Husbandry System on the Grass Intake and Egg Nutritive Characteristics of Laying Hens." *Journal of the Science of Food and Agriculture* 94(3, 2014): 459–67. Epub ahead of print, July 16, 2013.

Patzke, N., et al. "Consequences of Different Housing Conditions on Brain Morphology in Laying Hens." *Journal of Chemical Neuroanatomy* 37(2009): 141–48.

Sherwin, C. M., G. J. Richards, and C. J. Nicol. "Comparison of the Welfare of Layer Hens in Four Housing Systems in the UK." *British Poultry Science* 51(2010): 488–99.

Shimmura, T., et al. "Multi-factorial Investigation of Various Housing Systems for Laying Hens." *British Poultry Science* 51(2010): 31–42.

Shini, S. "Physiological Responses of Laying Hens to the Alternative Housing Systems." *International Journal of Poultry Science* 2(2003): 357–60.

Vallaeys, Charlotte, et al. *Scrambled Eggs: Separating Factory Farm Egg Production from Authentic Organic Agriculture*. Cornucopia, WI: Cornucopia Institute, 2010.

The Physiology of Stress

McEwen, B. S. "Stressed or Stressed Out: What Is the Difference?" *Journal of Psychiatry and Neuroscience* 30(5, 2005): 315–18.

Sapolsky, Robert. *Why Zebras Don't Get Ulcers*. New York: Holt Paperbacks, 2004.

Poultry Welfare and the Health Effects of Chronic Stress

Dawkins, M. S. "Behaviour as a Tool in the Assessment of Animal Welfare." *Zoology* 106(4, 2003): 383–87.

———. "A User's Guide to Animal Welfare Science." *Trends in Ecology and Evolution* 21(2, 2006): 77–82.

Huff, G. R., et al. "The Effect of Vitamin D_3 on Resistance to Stress-Related Infection in an Experimental Model of Turkey Osteomyelitis Complex." *Poultry Science* 79(5, 2000): 672–79.

Huff, G. R., et al. "Effect of Early Handling of Turkey Poults on Later Responses to Multiple Dexamethasone-Escherichia Coli Challenge. 2. Resistance to Air Sacculitis and Turkey Osteomyelitis Complex." *Poultry Science* 80(9, 2001): 1314–22.

Huff, G., et al. "Stress-Induced Colibacillosis and Turkey Osteomyelitis Complex in Turkeys Selected for Increased Body Weight." *Poultry Science* 85(2006): 266–72.

Shimmura, T., et al. "Overall Welfare Assessment of Laying Hens: Comparing Science-Based, Environment-Based, and Animal-Based Assessments." *Animal Science Journal* 82(2011): 150–60.

The U-Curve of Stress

Juster, R. P., et al. "A Transdisciplinary Perspective of Chronic Stress in Relation to Psychopathology Throughout Life Span Development." *Development and Psychopathology* 23(3, 2011): 725–76.

McEwen, B. S. "Physiology and Neurobiology of Stress and Adaptation: Central Role of the Brain." *Physiology Reviews* 87(2007): 873–904.

The Productivity Paradox

Chen, B. L., K. L. Haith, and B. Mullens. "A Beak Condition Drives Abundance and Grooming-Mediated Competitive Asymmetry in a Poultry Ectoparasite Community." *Parasitology* 138(2011): 748–57.

Just, N., C. Duchaine, and B. Singh. "An Aerobiological Perspective of Dust in Cage-Housed and Floor-Housed Poultry Operations." *Journal of Occupational Medicine and Toxicology* 4(2009): 13.

Karsten, H. D., et al. "Vitamins A, E, and Fatty Acid Composition of the Eggs of Caged Hens and Pastured Hens." *Renewable Agriculture and Food Systems* 25(2010): 45–54.

Lay, D. C., et al. "Hen Welfare in Different Housing Systems." *Poultry Science* 90(2011): 278–94.

Mench, J., D. Sumner, and J. T. Rosen-Molina. "Sustainability of Egg Production in the United States: The Policy and Market Context." *Poultry Science* 90(1, 2011): 229–40.

Mother Earth News. Egg vitamin D content (unpublished raw data). *Mother Earth News*, 2008.

Rimac, D., et al. "Exposure to Poultry Dust and Health Effects in Poultry Workers: Impact of Mould and Mite Allergens." *International Archives of Occupational and Environmental Health* 83(1, 2010): 9–19.

Singh, R., K. M. Cheng, and F. G. Silversides. "Production Performance and Egg Quality of Four Strains of Laying Hens Kept in Conventional Cages and Floor Pens." *Poultry Science* 88(2, 2009): 256–64.

Vallaeys, Charlotte, et al. *Scrambled Eggs: Separating Factory Farm Egg Production from Authentic Organic Agriculture*. Cornucopia, WI: Cornucopia Institute, 2010.

Human Productivity and Stress: The Corporate Perspective

De Jonge, J., et al. "Job Strain, Effort-Reward Imbalance and Employee Well-being: A Large-Scale Cross-Sectional Study." *Social Science and Medicine* 50(9, 2000): 1317–27.

Harter, James K., Frank L. Schmidt, and Corey L. M. Keyes. "Well-being in the Workplace and Its Relationship to Business Outcomes: A Review of the Gallup Studies." In *Flourishing: The Positive Person and the Good Life*. Washington, DC: American Psychological Association, 2003.

Shi, Y., et al. "Classification of Individual Well-being Scores for the Determination of Adverse Health and Productivity Outcomes in Employee Populations." *Population Health Management*, 2012. Available at: http://www.ncbi.nlm.nih.gov/pubmed/23013034 (accessed October 11, 2012).

Shimazu, A., et al. "Do Workaholism and Work Engagement Predict Employee Well-being and Performance in Opposite Directions?" *Industrial Health* 50(4, 2012): 316–21.

Promoting Plasticity

Battista, M. C., et al. "Intergenerational Cycle of Obesity and Diabetes: How Can We Reduce the Burdens of These Conditions on the Health of Future Generations?" *Experimental Diabetes Research* 2011(2011): 1–19.

Davidson, R. J., and B. S. McEwen. "Social Influences on Neuroplasticity: Stress and Interventions to Promote Well-being." *Nature Neuroscience* 15(5, 2012): 689–95.

Dobbs, David. "The Science of Success." *The Atlantic* (December 2009).

Green, M. K., et al. "Prenatal Stress Induces Long-Term Stress Vulnerability, Compromising Stress Response Systems in the Brain and Impairing Extinction of Conditioned Fear After Adult Stress." *Neuroscience* 192(2011): 438–51.

Lindqvist, C., et al. "Transmission of Stress-Induced Learning Impairment and Associated Brain Gene Expression from Parents to Offspring in Chickens." *PLOS ONE* 2(4, 2007): e364. Available at: doi:10.1371/journal.pone.0000364.

McEwen, B. S. "Stress, Sex, and Neural Adaptation to a Changing Environment: Mechanisms of Neuronal Remodeling." *Annals of the New York Academy of Sciences* 1204(2010): E38–59.

McEwen, B. S., and P. J. Gianaros. "Stress- and Allostasis-Induced Brain Plasticity." *Annual Review of Medicine* 62(2011): 431–45.

Vetencourt, J. F., et al. "The Antidepressant Fluoxetine Restores Plasticity in the Adult Visual Cortex." *Science* 320(2008): 385–88.

Alice Domar's Work on Stress and Infertility

Domar, Alice D., et al. "Original Report Relaxation Techniques for Reducing Pain and Anxiety During Screening." *American Journal of Roentgenology* 184(2005): 445–47.

Domar, A. D., et al. "Impact of a Group Mind/Body Intervention on Pregnancy Rates in IVF Patients." *Fertility and Sterility* 95(7, 2011): 2269–73.

Domar, Alice D., et al. "Lifestyle Behaviors in Women Undergoing In Vitro Fertilization: A Prospective Study." *Fertility and Sterility* 97(3, 2012): 697–701.

The Stress Reduction Toolbox

Gladwell, Malcolm. "The Roseto Mystery." In *Outliers: The Story of Success*. Boston: Little, Brown and Co., 2008.

Hawkley, L., and J. Cacioppo. "Loneliness and Pathways to Disease." *Brain, Behavior, and Immunity* 17(2003): S98–105.

Helliwell, J. F., and R. Putnam. "The Social Context of Well-being." *Philosophical Transactions of the Royal Society of London: Series B, Biological Sciences* 359(1449, 2012): 1435–46.

Kawachi, I., and L. F. Berkman. "Social Ties and Mental Health." *Journal of Urban Health: Bulletin of the New York Academy of Medicine* 78(3, 2001): 458–67.

Enhance Your Behavioral Freedom

Heimberg, R. G., et al. "Cognitive Behavioral Group Treatment for Social Phobia: Comparison with a Credible Placebo Control." *Cognitive Therapy and Research* 14(1, 1990): 1–23.

Hofmann, S., and J. Smits. "Cognitive-Behavioral Therapy for Adult Anxiety Disorders: A Meta-analysis of Randomized Placebo-Controlled Trials." *Journal of Clinical Psychiatry* 69(4, 2008): 621–32.

Shonkoff, J. P., W. T. Boyce, and B. S. McEwen. "Neuroscience, Molecular Biology, and the Childhood Roots of Health Disparities: Building a New Framework for Health Promotion and Disease Prevention." *Journal of the American Medical Association* 301(2009): 2252–59.

Play Tag and Dust-Bathe Often

Bavelier, D., et al. "Removing Brakes on Adult Brain Plasticity: From Molecular to Behavioral Interventions." *Journal of Neuroscience: The Official Journal of the Society for Neuroscience* 30(2010): 14964–71.

Brooks, A. W. "Get Excited: Reappraising Pre-Performance Anxiety as Excitement." *Journal of Experimental Psychology* 143(2013). Available at: http://www.ncbi.nlm.nih.gov/pubmed/24364682.

Juster, R. P., et al. "A Transdisciplinary Perspective of Chronic Stress in Relation to Psychopathology Throughout Life Span Development." *Development and Psychopathology* 23(2011): 725–76.

Get a Good Night's Roost

McEwen, B. S. "Central Effects of Stress Hormones in Health and Disease: Understanding the Protective and Damaging Effects of Stress and Stress Mediators." *European Journal of Pharmacology* 583(2008): 174–85.

Randall, David. *Dreamland: Adventures in the Strange Science of Sleep*. New York: W. W. Norton and Co., 2012.

Put the Right Foods in Your Gizzard

Appelhans, B. M., et al. "Depression Severity, Diet Quality, and Physical Activity in Women with Obesity and Depression." *Journal of the Academy of Nutrition and Dietetics* 112(5, 2012): 693–98.

Cohen, J., et al. "Psychological Distress Is Associated with Unhealthful Dietary Practices." *Journal of the American Dietetic Association* 102(5, 2002): 699–703.

Jacka, F. N., et al. "Association of Western and Traditional Diets with Depression and Anxiety in Women." *American Journal of Psychiatry* 167(3, 2010): 305–11.

Karsten, H. D., et al. "Vitamins A, E, and Fatty Acid Composition of the Eggs of Caged Hens and Pastured Hens." *Renewable Agriculture and Food Systems* 25(2010): 45–54.

McEwen, B. S. "Stress, Sex, and Neural Adaptation to a Changing Environment: Mechanisms of Neuronal Remodeling." *Annals of the New York Academy of Sciences* 1204(2010): E38–59.

Mugnai, C., et al. "The Effects of Husbandry System on the Grass Intake and Egg Nutritive Characteristics of Laying Hens." *Journal of the Science of Food and Agriculture* 94(3, 2014): 459–67. Epub ahead of print, July 16, 2013.

Omodei, D., and L. Fontana. "Calorie Restriction and Prevention of Age-Associated Chronic Disease." *FEBS Letters* 585(2011): 1537–42.

CHAPTER 4: SCRIBE WINERY

Beyond Whack-a-Mole Cancer Treatment

Begley, Sharon. "Rethinking the War on Cancer." *Newsweek,* September 5, 2008.

Drake, Nadia. "Forty Years on from Nixon's War, Cancer Research 'Evolves.' " *Nature Medicine* 17(7, 2011): 757.

Inadomi, J. M. "Surveillance in Barrett's Esophagus: A Failed Premise." *Keio Journal of Medicine* 58(1, 2009): 12–18.

Merlo, L.M.F., et al. "Cancer as an Evolutionary and Ecological Process." *Nature Reviews Cancer* 6(2006): 924–35.

Reid, B. J., R. Kostadinov, and C. Maley. "New Strategies in Barrett's Esophagus: Integrating Clonal Evolutionary Theory with Clinical Management." *Clinical Cancer Research* 17(11, 2011): 3512–19.

Quadratic Equations and Cabbage Moths

Chen, J., et al. "Solving the Puzzle of Metastasis: The Evolution of Cell Migration in Neoplasms." *PLOS ONE* 6(4, 2011): e17933. Available at: doi:10.1371/journal.pone.0017933.

Gatenby, R. A., and R. J. Gillies. "Why Do Cancers Have High Aerobic Glycolysis?" *Nature Reviews Cancer* 4(2004): 891–99.

Gatenby, R., J. Brown, and T. Vincent. "Lessons from Applied Ecology: Cancer Control Using an Evolutionary Double Bind." *Cancer Research* 69(2009): 7499–7502.

Gatenby, R. A. "A Change of Strategy in the War on Cancer." *Nature* 459(28, 2009): 508–9.

Gatenby, R. A., and R. J. Gillies. "Integrated Imaging of Cancer Metabolism." *Academic Radiology* 18(2011): 929–31.

Gillies, R. J., and R. A. Gatenby. "Hypoxia and Adaptive Landscapes in the Evolution of Carcinogenesis." *Cancer Metastasis Reviews* 26(2007): 311–17.

Laubenbacher, R., et al. "A Systems Biology View of Cancer." *Biochimica et Biophysica Acta* 1796(2, 2009): 129–39.

Silva, A. S., and R. Gatenby. "A Theoretical Quantitative Model for Evolution of Cancer Chemotherapy Resistance." *Biology Direct* 5(2010): 25.

The Farmer Who Thinks Like a Mountain

Campbell, Christy. *The Botanist and the Vintner: How Wine Was Saved for the World*. Chapel Hill, NC: Algonquin Books, 2006. This fascinating book offers a detailed account of the Phylloxera infestation.

Leopold, Aldo. "Thinking Like a Mountain." In *A Sand County Almanac*. New York: Ballantine Books, 1970.

IPM: Walking the Field

Abate, T., A. van Huis, and J.K.O. Ampofo. "Pest Management Strategies in Traditional Agriculture: An African Perspective." *Annual Review of Entomology* 45(2000): 631–59.

Birch, A.N.E., G. S. Begg, and G. R. Squire. "How Agro-ecological Research Helps to Address Food Security Issues Under New IPM and Pesticide Reduction Policies for Global Crop Production Systems." *Journal of Experimental Botany* 62(10, 2011): 3251–61.

Buhler, D. D. "50th Anniversary-Invited Article: Challenges and Opportunities for Integrated Weed Management." *Weed Science* 50(3, 2002): 273–80.

University of California. "Agriculture and Natural Resources: What Is Integrated Pest Management (IPM)?" Available at: http://www.ipm.ucdavis.edu/GENERAL/whatisipm.html (accessed October 25, 2012).

Monitoring Cancer: What Happens When You Start to Walk in the Field?

Caulin, A. F., and C. C. Maley. "Peto's Paradox: Evolution's Prescription for Cancer Prevention." *Trends in Ecology and Evolution* 26(2011): 175–82.

Maley, C. C., et al. "Genetic Clonal Diversity Predicts Progression to Esophageal Adenocarcinoma." *Nature Genetics* 38(2006): 468–73.

Merlo, L.M.F., et al. "A Comprehensive Survey of Clonal Diversity Measures in Barrett's Esophagus as Biomarkers of Progression to Esophageal Adenocarcinoma." *Cancer Prevention Research* 3(2010): 1388–97.

Reid, B. J. "Cancer Risk Assessment and Cancer Prevention: Promises and Challenges." *Cancer Prevention Research* 1(2008): 229–32.

Risques, R. A., et al. "Leukocyte Telomere Length Predicts Cancer Risk in Barrett's Esophagus." *Cancer Epidemiology, Biomarkers, and Prevention: A Publication of the American Association for Cancer Research* (cosponsored by the American Society of Preventive Oncology) 16(2007): 2649–55.

The Evo Game Master

Ackerman, Jennifer. "Food: How Altered?" *National Geographic* (May 2002).

Callahan, Rick. "Major Pest May Be Resistant to Genetically Modified Corn." *Huffington Post*, December 28, 2011.

Fungicide Resistance Action Committee (FRAC). "FRAC Code List: Fungicides Sorted by Mode of Action (Including FRAC Code Numbering).: Available at: http://www.frac.info/frac/publication/anhang/FRAC-Code-List2011-final.pdf (accessed January 31, 2012).

Stolte, D. "Trouble on the Horizon for Genetically Modified Crops?" *Science Daily*, June 19, 2012.

Boosting Beneficials

Liu, R. H. "Potential Synergy of Phytochemicals in Cancer Prevention: Mechanism of Action." *Journal of Nutrition* (suppl., 2004): 3479–85.

Maynard, M., et al. "Fruit, Vegetables, and Antioxidants in Childhood and Risk of Adult Cancer: The Boyd Orr Cohort." *Journal of Epidemiology and Community Health* 57(2003): 218–25.

Welch, A. A., et al. "Urine pH Is an Indicator of Dietary Acid-Base Load, Fruit and Vegetables and Meat Intakes: Results from the European Prospective Investigation into Cancer and Nutrition (EPIC)–Norfolk Population Study." *British Journal of Nutrition* 99(6, 2008): 1335–43.

Pheromone Traps and Sucker Gambits

Antonia, S. J., et al. "Combination of p53 Cancer Vaccine with Chemotherapy in Patients with Extensive Stage Small Cell Lung Cancer." *Clinical Cancer Research: An Official Journal of the American Association for Cancer Research* 12(2006): 878–87.

Buonaguro, L., et al. "Systems Biology Applied to Vaccine and Immunotherapy Development." *BMC Systems Biology* 5(2011): 146.

Leaf-Stripping and Environmental Control

Alexandre, L., et al. "Systematic Review: Potential Preventive Effects of Statins Against Oesophageal Adenocarcinoma." *Alimentary Pharmacology and Therapeutics* 36(2012): 301–11.

Hawkley, L. C., and J. T. Cacioppo. "Loneliness and Pathways to Disease." *Brain, Behavior, and Immunity* 17(2003): S98–105.

Madden, K. S., M. J. Szpunar, and E. B. Brown. "Early Impact of Social Isolation and Breast Tumor Progression in Mice." *Brain, Behavior, and Immunity* (2012). Available at: http://dx.doi.org/10.1016/j.bbi.2012.05.003.

Maley, C. C., B. J. Reid, and S. Forrest. "Cancer Prevention Strategies That Address the Evolutionary Dynamics of Neoplastic Cells: Simulating Benign Cell Boosters and Selection for Chemosensitivity." *Cancer Epidemiology, Biomarkers, and Prevention* 13(2004): 1375–84.

Omodei, D., and L. Fontana. "Calorie Restriction and Prevention of Age-Associated Chronic Disease." *FEBS Letters* 585(2011): 1537–42.

Vaughan, T. L., et al. "Nonsteroidal Anti-inflammatory Drug Use, Body Mass Index, and Anthropometry in Relation to Genetic and Flow Cytometric Abnormalities in Barrett's Esophagus." *Cancer Epidemiology, Biomarkers, and Prevention* 11(2002): 745–52.

Tailoring the Treatment to the Field

Gatenby, R. A., et al. "Adaptive Therapy." *Cancer Research* 69(2009): 4894–4903.

Wolves, Snakes, and Barrett's

Reid, B. J., R. Kostadinov, and C. C. Maley. "New Strategies in Barrett's Esophagus: Integrating Clonal Evolutionary Theory with Clinical Management." *Clinical Cancer Research: An Official Journal of the American Association for Cancer Research* 17(2011): 3512–19.

A New Take on Cancer

Mukherjee, Siddhartha. *The Emperor of All Maladies: A Biography of Cancer.* New York: Scribner, 2010. This is a wonderful biography of this misunderstood disease.

CHAPTER 5: LA FAMILIA VERDE URBAN FARMS

parentearth. "Food Hero: Karen Washington." YouTube, October 30, 2012.

Plsek, P. E., and T. Greenhalgh. "Complexity Science: The Challenge of Complexity in Health Care." *British Medical Journal* 323(7313, 2001): 625–28.

Food Deserts and Food Mirages

Boone-Heinonen, J., et al. "Fast Food Restaurants and Food Stores: Longitudinal Associations with Diet in Young to Middle-aged Adults: The CARDIA Study." *Archives of Internal Medicine* 171(13, 2011): 1162–70.

Croft, Cammie. "Taking on 'Food Deserts.' " The White House Blog, February 24, 2010. Available at: http://www.whitehouse.gov/blog/2010/02/24/taking-food-deserts.

Cummins, S., et al. "Large-Scale Food Retailing as an Intervention for Diet and Health: Quasi-experimental Evaluation of a Natural Experiment." *Journal of Epidemiology and Community Health* 59(2005): 1035–40.

Fielding, J. E., and P. A. Simon. "Food Deserts or Food Swamps?" *Archives of Internal Medicine* 171(13, 2011): 1171–72.

Kohan, Eddie Gehman. "Combating Food Deserts, Creating Jobs." Obama Foodorama, July 21, 2011. Available at: http://obamafoodorama.blogspot.com/2011/07/michelle-obama-our-lady-of-food-deserts.html (accessed October 25, 2012).

Kropf, M. L., et al. "Food Security Status and Produce Intake and Behaviors of Special Supplemental Nutrition Program for Women, Infants, and Children and Farmers' Market Nutrition Program Participants." *Journal of the American Dietetic Association* 107(2007): 1903–8.

Urban Farming: A Public Health Intervention?

Alaimo, K., et al. "Fruit and Vegetable Intake Among Urban Community Gardeners." *Journal of Nutrition Education and Behavior* 40(2, 2008): 94–101.

Brown, K. H., and A. L. Jameton. "Public Health Implications of Urban Agriculture." *Journal of Public Health Policy* 21(1, 2000): 20–39.

Hale, J., et al. "Connecting Food Environments and Health Through the Relational Nature of Aesthetics: Gaining Insight Through the Community Gardening Experience." *Social Science and Medicine* 72(11, 2011): 1853–63.

Litt, J. S., et al. "The Influence of Social Involvement, Neighborhood Aesthetics, and Community Garden Participation on Fruit and Vegetable Consumption." *American Journal of Public Health* 101(2011): 1466–73.

McCormack, L. A., et al. "Review of the Nutritional Implications of Farmers' Markets and Community Gardens: A Call for Evaluation and Research Efforts." *Journal of the American Dietetic Association* 110(2010): 399–408.

Rodrigues, K., and J. Litt. "Plant a Row, Grow a Community: Why Gardens Make Sense for Public Health." Unpublished paper, October 25, 2012.

Wakefield, S., et al. "Growing Urban Health: Community Gardening in South-East Toronto." *Health Promotion International* 22(2007): 92–101.

Weltin, A. M., and Lavin, R. P. "The Effect of a Community Garden on HgA1c in Diabetics of Marshallese Descent." *Journal of Community Health Nursing* 29(2012): 12–24.

The Farmstand Factor

Racine, E. F., Smith Vaughn, A., and Laditka, S. B. "Farmers' Market Use Among African-American Women Participating in the Special Supplemental Nutrition Program for Women, Infants, and Children." *Journal of the American Dietetic Association* 110(2010): 441–46.

McCormack, L. A., Laska, M. N., Larson, N. I., and Story, M. "Review of the Nutritional Implications of Farmers' Markets and Community Gardens: A Call for Evaluation and Research Efforts." *Journal of the American Dietetic Association* 110(2010): 399–408.

Digging for Longevity

Van den Berg, A. E., van Winsum-Westra, M., de Vries, S., and van Dillen, S. M. E. "Allotment Gardening and Health: A Comparative Survey Among Allotment Gardeners and Their Neighbors Without an Allotment." *Environmental Health: A Global Access Science Source* 9(2010): 74.

Verghese, J. et al. "Leisure Activities and the Risk of Dementia in the Elderly." *The New England Journal of Medicine* 348(2003): 2508–16.

Vermeulen, J. et al. "Does a Falling Level of Activity Predict Disability Development in Community-Dwelling Elderly People?" *Clinical Rehabilitation* (2012).

Contagious Vegetable Eating

Heim, S., Stang, J., and Ireland, M. "A Garden Pilot Project Enhances Fruit and Vegetable Consumption Among Children." *Journal of the American Dietetic Association* 109(2009): 1220–26.

Carney, P. et al. "Impact of a Community Gardening Project on Vegetable Intake, Food Security and Family Relationships: A Community-Based Participatory Research Study." *Journal of Community Health* 37(2012): 874–81.

Erinosho, T. O., Dixon, L. B., Young, C., Brotman, L. M., and Hayman, L. L. "Caregiver Food Behaviours Are Associated with Dietary Intakes of Children Outside the Child-Care Setting." *Public Health Nutrition* 1–10 (2012).

Urban Farming: A Hothouse for Community Enterprise

See Policy Link's report on the connections between growing food, growing jobs, and improving public health: http://www.policylink.org/atf/cf/%7B97C6D565-BB43-406D-A6D5-ECA3BBF35AF0%7D/URBAN%20AG_FULLREPORT_WEB1.PDF.

More Vegetables, Less Crime—What's the Link?

Helliwell, J. F., and R. D. Putnam. "The Social Context of Well-being." *Philosophical Transactions of the Royal Society of London: Series B, Biological Sciences* 359(2004): 1435–46.

Kropf, M. L., et al. "Food Security Status and Produce Intake and Behaviors of Special Supplemental Nutrition Program for Women, Infants, and Children and Farmers' Market Nutrition Program Participants." *Journal of the American Dietetic Association* 107(2007): 1903–8.

Sampson, R. J., S. W. Raudenbush, and F. Earls. "Neighborhoods and Violent Crime: A Multilevel Study of Collective Efficacy." *Science* 277(1997): 918–24.

Teig, E., et al. "Collective Efficacy in Denver, Colorado: Strengthening Neighborhoods and Health Through Community Gardens." *Health and Place* 15(2009): 1115–22.

To Be Healthy, You Have to Love the Place Where You Live

Hale, J., et al. "Connecting Food Environments and Health Through the Relational Nature of Aesthetics: Gaining Insight Through the Community Gardening Experience." *Social Science and Medicine* 72(11, 2011): 1853–63.

Morland, K., S. Wing, and A. D. Roux. "The Contextual Effect of the Local Food Environment on Residents' Diets: The Atherosclerosis Risk in Communities Study." *American Journal of Public Health* 92(11, 2002): 1761–67.

Edible Schoolyards

Hoffman, J. "Q&A: Education from the Ground Up." *Nature* 459(2009): 913.

Rauzon, S., et al. *An Evaluation of the School Lunch*. Berkeley: UC Berkeley, 2010. Available at: http://edibleschoolyard.org/sites/default/files/file/An_Evaluation_of_the_School_Lunch_Initiative_Final%20Report_9_22_10.pdf (accessed October 25, 2012).

CHAPTER 6: MORNING MYST

Harth, W., K. Seikowski, and B. Hermes. "Lifestyle Drugs in Old Age—A Mini-Review." *Gerontology* 55(1, 2009): 13–20. Epub ahead of print, November 12, 2008.

Misadventures in Beautyland

Bensouilah, Janetta, and Philippa Buck. *Aromadermatology: Aromatherapy in the Treatment and Care of Common Skin Conditions*. Boston: Blackwell Publishers, 2006.

Viriditas

Catty, Suzanne. *Hydrosols: The Next Aromatherapy*. Rochester, NY: Healing Arts Press, 2001.

Hildegard of Bingen. *Selected Writings: Hildegard of Bingen*. Translated by Mark Atherton. New York: Penguin, 2001.

Rose, Jeanne. "Hydrosols." The Aromatic Plant Project. Available at: http://www.aromaticplantproject.com/hydrosols.html (accessed October 25, 2012). This is a good website to learn more about hydrosols.

The Human Diffuser

Brennan, P., and K. M. Kendrick. "Mammalian Social Odours: Attraction and Individual Recognition." *Philosophical Transactions of the Royal Society of London: Series B, Biological Sciences* 361(2006): 2061–78.

Gelstein, S., et al. "Human Tears Contain a Chemosignal." *Science* 331(2011): 226–30.

Grammer, K., B. Fink, and N. Neave. "Human Pheromones and Sexual Attraction." *European Journal of Obstetrics and Gynecology* 118(2005): 135–42.

Hoover, K. C. "Smell with Inspiration: The Evolutionary Significance of Olfaction." *American Journal of Physical Anthropology* 143(suppl., 2010): 63–74.

Jacobs, L. F. "From Chemotaxis to the Cognitive Map: The Function of Olfaction." *Proceedings of the National Academy of Sciences of the United States of America* 109(suppl., 2012): 10693–700.

Pause, B. M. "Processing of Body Odor Signals by the Human Brain." *Chemosensory Perception* 5(2012): 55–63.

Schnaubelt, Kurt. *Medical Aromatherapy: Healing with Essential Oils*. Berkeley, CA: North Atlantic Books, 1999.

More Reading on Deep Homology

Dawkins, Richard. *The Selfish Gene*, 30th anniversary edition. New York: Oxford University Press, 2006.

Gould, Stephen Jay. *The Structure of Evolutionary Theory*. Cambridge, MA: Harvard University Press, 2002. This book offers an extensive analysis of deep homology and parallel evolution.

Pertseva, M. N., and A. O. Shpakov. "The Prokaryotic Origin and Evolution of Eukaryotic Chemosignaling Systems." *Neuroscience and Behavioral Physiology* 39(8, 2009): 793–804.

Shubin, N., C. Tabin, and S. Carroll. "Deep Homology and the Origins of Evolutionary Novelty." *Nature* 457(2009): 818–23.

Plant-Human Miscommunication

Lichtenberg, E. "The Economics of Cosmetic Pesticide Use." *American Journal of Agricultural Economics* 79(1, 1997): 39–46.

Pimentel, D., et al. "Pesticides, Insects in Foods, and Cosmetic Standards." *BioScience* 27(3, 1977): 178–85.

Thompson, G. D., and J. Kidwell. "Explaining the Choice of Organic Produce: Cosmetic Defects, Prices, and Consumer Preferences." *American Journal of Agricultural Economics* 80(2, 1998): 277–87.

U.S. Department of Agriculture. "United States Standards for Grades of Apples." December 19, 2002. Available at: http://www.ams.usda.gov/AMSv1.0/getfile?dDocName=STELPRDC5050339.

———. "National Organic Program: Cosmetics, Body Care Products, and Personal Care Products." 2008. Available at: http://www.ams.usda.gov/AMSv1.0/getfile?dDocName=STELPRDC5068442 (accessed October 30, 2012).

Consult the Genius of the Place

Andrade-Cetto, A., and M. Heinrich. "Mexican Plants with Hypoglycaemic Effect Used in the Treatment of Diabetes." *Journal of Ethnopharmacology* 99(3, 2005): 325–48.

Roundell, Mrs. Charles. *The Still-Room*. London: Bodley Head, 1903.

CONCLUSION: FIVE "AHAS!" AND ONE SUGGESTION

Leopold, Aldo. *Round River.* New York: Oxford University Press. 1993.

Goldberger, A. "Non-linear Dynamics for Clinicians: Chaos Theory, Fractals, and Complexity at the Bedside." *Lancet* 347(1996): 1312–14.

Index